Marangoni and Interfacial Phenomena in Materials Processing

Originating from contributions to a
Discussion of the Royal Society of London

Organized and Edited by
E. D. Hondros, M. McLean
and K. C. Mills

Book 692
Published in 1998 by
IOM Communications
in association with
The Royal Society

All papers originally appeared in
*Philosophical Transactions of the
Royal Society of London*
A **356**, 811–1061

Volume ©1998 The Royal Society
Papers by Olson & Edwards and
Mills *et al.* are British Crown Copyright
All rights reserved

ISBN 1-86125-056-8

IOM Communications Ltd
1 Carlton House Terrace
London SW1Y 5DB

IOM Communications Ltd
is a wholly owned subsidiary of
The Institute of Materials

Typeset in the UK by
T&T Productions Ltd
Printed and bound in the UK at
The University Press, Cambridge

Contents

Preface

Convection in fluids associated with gradients in surface tension, whether associated with temperature or concentration gradients, has been studied for over a century. There has been a growing awareness of their significance to a wide range of materials processes. This Meeting has brought together leading scientists who are elucidating the fundamental aspects of the Marangoni effect, with those with an industrial interest in exploiting the Marangoni effect to optimize material processing in diverse industrial sectors. The significance of the effect in liquid metal processing (such as in steel making, welding, coating, secondary melting) is shown to be particularly strong and to be a major factor in guiding the approaches to industrial control of these processes. It is hoped that these proceedings will stimulate interest in applying these concepts to other processes where they undoubtedly play an important role.

E. D. HONDROS
K. C. MILLS
M. MCLEAN

Introduction: significance of capillary driven flows in materials processing

By E. D. Hondros

Department of Materials, Imperial College of Science, Technology and Medicine, London SW7 2BZ, UK

In this introductory paper to the Royal Society Discussion Meeting, Marangoni and Interfacial Phenomena in Materials Processing, we present first the historical background to the general phenomenon of capillary gradient-driven flows, or 'Marangoni' flows. The many early observations in organic and inorganic liquids reflect the intense scientific curiosity surrounding this and other capillarity-induced phenomena. However, only in recent decades has the significance of this effect in various industrial technologies been appreciated. In this volume, we have deliberately focused on Marangoni observations relevant to the procedures and technologies involved in the production and processing of materials. The papers collected here demonstrate a clear significance of Marangoni flows in a variety of processes.

Keywords: Marangoni flow; high temperature materials; materials processing; capillary action; surface tension gradients

---

> '... numerous motions of extremely curious and wonderful characters in fluids undergoing evaporation.'
>
> James Thomson 1855

The broad theme of this volume relates closely to the science of capillary action in fluids and, in particular, how one somewhat neglected aspect of this science is seen today to bear on important phenomena associated with the processing of modern materials. Since capillarity has been one of the important classical branches of study in the natural sciences, it is no surprise that nearly all the famous names of science of the 18th and 19th centuries were attracted to this field, including, among many others, Gauss, Young, Franklin, Laplace and Raleigh. Thus much of the basic understanding of this theme had become established by the turn of the 20th century; see, for example, the seminal article, 'Capillary action', by Maxwell (1878), who also describes the historical development of the subject up to that time.

Specifically, we are concerned here with phenomena induced by surface tension gradients which derive from temperature or surface concentration variations. During the middle of the last century, James Thomson (1855), the elder brother of Lord Kelvin, experimented with the spreading of alcohol drops on the surface of water and was probably the first to invoke the idea of surface tension-driven flows to explain this and many other intriguing manifestations of capillary gradients, such as the 'tears of strong wine' in liquids containing alcohol, or the bizarre 'camphor dance'. The extraordinary spread of interest in such surface phenomena in the past century and the resulting comprehensive literature on the subject (Tomlinson (1873) cites 37 references in his paper on the motions of camphor) surely reflects the easy accessibility of such study, requiring only elementary chemical substances and equipment, accurate observations and an enquiring intellect; in the absence of a bewildering armoury of the sophisticated equipment of our time, anyone could dabble in the intricacies of soap bubbles.

Without entering into details of the numerous fascinating studies involving movements on liquid surfaces, and the heatedly contested claims on the priority of discovery, it suffices for our purposes to mention the publication by Marangoni (1878), who provided a wealth of detailed information on the effects of variations of the potential energy of liquid surfaces arising from variations in temperature and composition. These effects have now come to be conveniently associated with his name, noting incidentally that Thomson's (1855) short but illuminating paper was curiously forgotten during these debates on priority.

Among the more important general phenomena involving Marangoni flows, we note that associated with the name of Bénard (1901), which refers to the formation of a polygonal cellular structure in a thin liquid layer heated from below, an effect discovered recently in thin semiconductor films, and a subject of continual interest and theoretical analysis, as reported in this volume.

Yet for decades, the earlier numerous observations lay in the realm of interesting scientific curiosities, useful for illustrating the wonders of surface tension. The role of the basic effect in technology was probably first demonstrated by chemical engineers in the field of liquid–liquid extraction: there is a natural progression from classical observations of the twitching of air bubbles in alcohol containing water, to the spontaneous agitation of interfaces between unequilibrated fluids, leading to the contemporary extensive knowledge in the field of interfacial turbulence.

This Discussion Meeting is firmly in the field of materials processing, encompassing all the procedures whereby raw products are transformed into useful engineering materials: arguably the study of materials processing is one of the high priority R&D pursuits in most materials research centres today.

We focus here on a variety of observations in materials processing, particularly in metallic systems which have been suspected to demonstrate Marangoni flows: e.g. in pyrometallurgy, in which interest lies in interfacial mass transfer; in the complex mixing hydrodynamics of furnaces containing melts; in the erosion of the walls of ceramic crucibles containing liquid metals. These instances reflect both a scientific and industrial interest. In fact, phenomena attributable to Marangoni flows have been reported in innumerable instances relevant to modern technologies, such as in hot salt corrosion in aero-turbine blades; the drying of solvent-containing paints; the drying of silicon wafers used in electronics; even the modelling of the role of Marangoni flows in the micro-pools produced by plasma disruptions on the first wall of a prospective thermonuclear fusion reactor.

We note the disparate nature of the above observations, the only link between them being the suspected or demonstrated presence of Marangoni flows. Each of the subjects has been explored separately by workers pursuing problems in their own field, whether this be weld penetration or crucible erosion; and in each subject area, those concerned have arrived, often with some hesitation, at an explanation based on capillary-driven flows. Hence, one of our objectives in this meeting has been to stimulate interactions between theoreticians and specialists engaged in the various subject areas. Indeed, in order to understand the various phenomena in the materials processing field, the simple classical treatment has in general been sufficient. The urgent requirement has been for reliable high temperature physicochemical data, such as surface tension, viscosity and thermal conductivity in alloy systems, a matter which is now beginning to be seriously addressed and is reported in this volume. Again, the recent theoretical studies of more complex convectional patterns, such as those reported here in Bénard cellular systems, should be valuable in future studies of high temperature multiphase metallic melts.

There is one contemporary area in which Marangoni phenomena have been studied in a coordinated manner, namely in melt and crystal processes in microgravity conditions, associated with space research programmes. Here, a whole literature has now emerged, following the considerable experimental and theoretical research carried out which is reported in regular conferences devoted to the broader field of materials and fluidic processes in microgravity. We note that leading workers from this field have contributed to this Meeting and also in a one-day associated event sponsored by the CIBA Foundation. Their experience will stimulate these Discussions on materials processes important to industry. One of the prime needs of our theme is cross-fertilization.

In this Meeting, which started off as a tentative exploration of past speculative reports, we have been rewarded by several papers which, through dedicated experimentation and the application of new data on physicochemical properties combined with computer modelling, not only confirm earlier speculations, but also reveal original insights and unexpected Marangoni-related phenomena, in particular, in materials technologies.

As examples, we mention the following:

(1) The confirmation that the problem of 'flux line erosion', which has for years bedeviled the glass-making and pyrometallurgical industries, is clearly related to Marangoni flows which wash the refractory wall with a thin film of corrosive slag from the bulk. Here we are tempted to quote the final remarks of Professor Makai in his paper: 'The trivial phenomenon of "tears of wine" that was caught by the piercing eyes of the scientist Thomson, holds the key to the solution of the industrially serious problem of corrosion of refractories.'

(2) Again, by applying newly acquired data on the surface tension of alloys at high temperatures in combination with modelling of localized melting, there is a clear demonstration of how surface active solutes in steel can switch the temperature coefficient of surface tension to positive values; in turn, this reverses the direction of the Marangoni flows in the molten pool during arc or laser welding, giving rise to deeper penetration of molten metal.

(3) In another elegant example of computerized modelling, a team from the NPL and Imperial College has demonstrated that in the rapidly growing technology of producing clean industrial alloys for advanced applications using electron beam melting, Marangoni flows driven by surface active additives can be used to control the distribution and removal of inclusions.

(4) In a more theoretical description backed up by experimental data, a team from Grenoble shows that in the reactive spreading of liquid alloy droplets over a ceramic substrate (which bears on the modern technology of multi-material processing by liquid metal infiltration), Marangoni convection governs the process kinetics through the transport of reactive species to the spreading front.

(5) Some of the problems of growth of both organic and inorganic crystals in microgravity relate to perturbations arising from Marangoni flows, problems that are being tackled by the microgravity research community. Of more general interest, a Japanese team reports that in the growth of silicon crystals from the melt in micro-

gravity conditions, Marangoni instabilities abound, through temperature differences at the solid–liquid interfaces and also through the extreme sensitivity of the surface tension of silicon to oxygen in the ambient atmosphere.

Finally, it is hoped that this volume will help to extend to a wider audience the subject of Marangoni hydrodynamics and an appreciation of their relevance to materials phenomena; and furthermore, as in other analytical approaches used in modern materials research, the knowledge of capillary action and capillary-induced flow might form part of the common scientific equipment of students and workers engaged in this growing and commercially rewarding field of materials processing.

References

Bénard, H. 1901 Les tourbillons cellulaires dans une nappe liquide transportant de la chaleur par convection on régime permanent. *Ann. Chim. Phys.* Ser. 7, **23**, 62–144.

Marangoni, C. 1878 Difesa della teoria dell'elasticità superficiace dei liquidi. *Nuovo Cim.* Ser. 3, **3**, 97–115.

Maxwell, G. J. C. 1878 Capillary action. *Encyclopaedia Brittanica*, 9th edn. New York.

Thomson, J. 1855 On certain curious motions observable at the surfaces of wine and other alcoholic liquors. *Phil. Mag.* Ser. 4, **10**, 330–333.

Tomlinson, C. 1873 On the motion of camphor and of certain liquids on the surface of water. *Phil. Mag.* **46**, 376–388.

Influence of capillarity on chemical stability and of electric field on surface tension near the critical point

BY A. SANFELD

Laboratoire de Modélisation en Mécanique et Thermodynamique (LMMT),
UPRES 193, Université d'Aix-Marseille, Faculté des Sciences de St Jérôme Bd,
Escadrille Normandie Niemen, 13397 Marseille Cédex 20, France

In the present paper, we show that the stability of a chemical reaction occurring in a capillary layer depends on surface properties. A chemical reaction unstable in a bulk phase may be stabilized in a surface layer and, vice versa, a stable reaction in a bulk phase may be unstable when occurring at an interface. The second part of this paper shows how an applied electric field may be responsible for a shift of the critical point through the effect on surface tension.

Keywords: surface tension; stability; electric field; critical point

1. Introduction

For the physicist and the chemist, one of the primary interests in thermodynamics lies in its ability to establish a criterion of the stability of a given chemical or physical transformation under specified conditions. The first law of thermodynamics describes the conservation of energy while the second law expresses the evolution of all irreversible processes. The formulation of these two universal principles is, however, unable to forecast a stability transition for equilibrium or far from equilibrium states. The stability analysis gives the conditions for the regression or the amplification of the fluctuations of any physical and chemical variable.

It is well known that spontaneous microscopic fluctuations always occur in all systems. When these fluctuations are damped in time, and do not grow to change the macroscopic state, the system is stable. When dealing with stability problems, three types of situation have to be considered: the equilibrium states; the linear region near equilibrium; and the non-equilibrium situations beyond the linear region. At equilibrium, far from a phase transition, and near equilibrium, fluctuations play a minor role. They give 'corrections' to macroscopic results. In the region of phase transition, equilibrium instabilities give rise to the formation of equilibrium structures that are maintained in reversible processes, or for processes that slightly remove the system from equilibrium. To study the equilibrium stability, the concept of thermodynamic potentials was first used, such as the Helmholtz free energy F, or other potentials (Gibbs 1928), according to the boundary conditions. Later, a concept of stability which was not dependent on the boundary conditions was introduced (De Donder 1942). The method adopted by Prigogine & Defay (1967) is based upon the direct evaluation of the entropy production in the course of a perturbation and so permits a discussion of stability with respect to any kind of perturbation.

In his famous 'Traité d'Energétique', Duhem (1911, ch. XVI) used the basic idea of Gibbs to obtain thermal and mechanical conditions for the stability of equilibrium states. They derived explicitly the curvature of the specific mass entropy $\delta^2 s$ ($\delta^2 s \leqslant 0$ for stable states) in a quadratic form containing fluctuations of temperature and density. Later, Glansdorff & Prigogine (1970) extended their theory to diffusion or chemical stability and Sanfeld & Steinchen (1971) to charged layers. Far from equilibrium the situation is totally different. In that case, fluctuations play a major role by changing the spatio-temporal symmetry of the system. However, contrary to equilibrium transitions, new types of transitions can give rise to a new order characterized by a macroscopic spatial scale. Such states, called by Glansdorff–Prigogine 'dissipative structures', are able to exhibit a permanent temporal activity, regular or chaotic with a macroscopic time scale. Considering that external or internal perturbations (amplification of fluctuations) can destabilize the state of a system, two cases are generally discussed: the steady states and the mobile boundaries. In the former case, the stability criterion (sufficient condition) is given by the product of the fluxes and generalized force fluctuations. In the latter case, a more general formulation based on the trajectories of the processes is needed. The basic idea was introduced and developed by Lyapunov (Cesari 1963; Minorsky 1962; Lin 1967) who introduced a quantity related to the initial perturbation at a given time and which was defined to be positive or negative. If the time derivative of this quantity $((\partial/\partial t)\delta^2 s)$ has the reverse sign of the same quantity, the stability condition for non-equilibrium states is satisfied. If we assume that the state of local equilibrium is stable, the fundamental hypothesis of local equilibrium $\delta^2 s$ can be used as a Lyapunov function. The choice of $\delta^2 s$, instead of any other quadratic function, finds its justification in its physical significance in terms of fluctuations in the Einstein theory (Landau & Lifshitz 1958). In the first part of our study we shall apply the Lyapunov concept of stability to a surface chemical reaction in order to display the role of the surface tension. In addition to the thermodynamic stability approach many authors have devoted more attention to hydrodynamical treatments (Chandrasekar 1961; Lin 1967; Drazin & Reid 1981). One of the major points is the role played by the surface tension on the stability of fluid surfaces and more specifically of external fields acting at gas–liquid or liquid–liquid interfaces (Sørensen 1979).

In § 3 we describe the influence of an applied electric field on the surface near the critical point (Gouin & Sanfeld 1997). As the surface tension is zero at the critical point, the interface vanishes. Therefore the influence of an applied external field in that region could be of importance to maintain a discontinuity (or semi) domain between the two bulk phases. To this purpose we use a generalized chemical potential (de Groot & Mazur 1962, p. 396; Sanfeld 1968) combined with a statistical mean field theory (Rowlinson & Widom 1982).

2. Non-equilibrium stability of a plane surface

As previously shown by Steinchen & Sanfeld (1981) for capillary systems, the local time derivative of a Lyapunov function including the curvature of entropy and a convective term contains boundary flux terms, Laplace, Marangoni and surface Benard effects and fluctuations of flux-forces. We shall restrict here our analysis to pure chemical reactions in a monolayer at constant and uniform temperature with no accumulation of matter at the sublayer. We shall then compare the monolayer phase to the volume phase. Both systems are considered at rest and for appropriate Newmann or Dirichlet boundary conditions. Restricted to small perturbations around a non-equilibrium process, a sufficient condition for the local stability of pure chemical reactions in a bulk phase (Glansdorff & Prigogine 1970) is given by

$$\frac{1}{2}\frac{\partial}{\partial t}\delta^2 s = T^{-1}\sum_\rho \delta\mathcal{A}_\rho \delta v_\rho \geqslant 0, \tag{2.1}$$

where $\mathcal{A}_\rho$ and v_ρ are the chemical affinity and rate of reaction ρ, respectively.

For reactions in a monolayer (superscript 'c') the stability criterion reads (Steinchen & Sanfeld 1981)

$$\frac{1}{2}\frac{\partial}{\partial t}\delta^2 s = T^{-1}\sum_\rho \delta\mathcal{A}_\rho^c \delta v_\rho^c \geqslant 0. \tag{2.2}$$

In order to calculate the chemical contribution $\delta\mathcal{A}_\rho^c \delta v_\rho^c$ in equation (2.2), let us now consider a totally irreversible surface chemical reaction ρ characterized by the general kinetic law

$$v_\rho^c = k_\rho^c (\gamma_\gamma^c)^{m_\gamma} (\gamma_\beta^c)^{m_\beta} \cdots, \tag{2.3}$$

where k_ρ^c is the kinetic constant; γ_γ^c, γ_β^c are the surface activities (in terms of surface concentrations) of γ; m_γ, m_β are the partial orders. Strictly speaking, k_ρ^c is a pseudo-constant. Indeed, k_ρ^c is directly related to the surface activation energy which depends on the state of the monolayer lattice, i.e. the capillary parameters. By neglecting the fluctuations of k_ρ^c, as a first approximation, we get, for only one fluctuating component β,

$$\delta v_\rho^c = k_\rho^c (\gamma_\gamma^c)^{m_\gamma} \cdots m_\beta (\gamma_\beta^c)^{m_\beta^{-1}} \delta\gamma_\beta^c. \tag{2.4}$$

Now we have to calculate the second factor $\delta\mathcal{A}_\rho^c$ given by

$$\delta\mathcal{A}_\rho^c = -\sum_\gamma \nu_{\rho\gamma}\delta\mu_\gamma^c, \tag{2.5}$$

where the chemical potential in the monolayer μ_γ^c (Defay & Prigogine 1966, p. 166) reads

$$\mu_\gamma^c = \mu_\gamma^\infty + RT\ln\gamma_\gamma^c - RT\ln\Gamma - \sigma a_\gamma, \tag{2.6}$$

where μ_γ^∞ is the standard chemical potential, Γ the dimensionless total surface molar concentration ($\Gamma = \sum\Gamma_\gamma$). The activity coefficient f_γ^c is defined by

$$\gamma_c^c = f_\gamma^c \Gamma_\gamma. \tag{2.7}$$

At constant T and p and for one fluctuating reactant β,

$$\delta\mathcal{A}_\rho^c = RT\left\{-\nu_{\rho\beta}\frac{\delta\gamma_\beta^c}{\gamma_\beta^c} + \nu_{\rho\beta}\frac{\delta\Gamma_\beta}{\Gamma} + \delta\left(\sum_\gamma \nu_{\rho\gamma}a_\gamma\sigma/RT\right)\right\}. \tag{2.8}$$

As $\Gamma > \Gamma_\beta$,

$$\frac{\delta\Gamma_\beta}{\Gamma} \ll \frac{\delta\Gamma_\beta}{\Gamma_\beta},$$

so that equation (2.8) reduces to

$$\delta\mathcal{A}_\rho^c = RT\left\{-\nu_{\rho\beta}\frac{\delta\gamma_\beta^c}{\gamma_\beta^c} + \delta\left(\sum_\gamma \nu_{\rho\gamma}a_\gamma\sigma/RT\right)\right\}. \tag{2.9}$$

Taking into account the fluctuations of the chemical rate (2.4), one has finally

$$\tfrac{1}{2}\partial_t(\delta^2 s^c)_\rho = T^{-1}\delta v_\rho^c \delta \mathcal{A}_\rho^c$$

$$= Rk_\rho^c(\gamma_\rho^c)^{m_\gamma}\cdots m_\beta(\gamma_\rho^c)^{m_{\beta}-1}\left\{-\nu_{\rho\beta}\frac{(\delta\gamma_\beta^c)^2}{\gamma_\beta^c} + \delta\left(\sum_\gamma \nu_{\rho\gamma} a_\gamma \sigma/RT\right)\delta\gamma_\beta^c\right\}.$$

$$(2.10)$$

For the same mechanism in the bulk phase, we will have

$$\tfrac{1}{2}\partial_t(\delta^2 s)_\rho = T^{-1}\delta v_\rho \delta \mathcal{A}_\rho = -k(\gamma_\gamma)^{m_\gamma}\cdots m_\beta(\gamma_\beta)^{m_{\beta}-2}\{R\nu_{\rho\beta}(\delta\gamma_\beta)^2\}, \qquad (2.11)$$

where the activity of component γ in the bulk phase is defined by $\gamma_\gamma = f_\gamma C_\gamma$. The autocatalytic (or non-autocatalytic) character of a bulk phase chemical reaction is characterized by the sign of $\nu_{\rho\beta}$. As shown by expression (2.10), there is an additional contribution in the monolayer due to the intrinsic contribution of the surface free energy to the chemical potential, i.e. $(a_\gamma \sigma)$. Suppose now a monolayer composed, either by reacting solutes of surfactant impurities and gaseous species on a liquid metal support, or by a non-reacting rare gaseous solvent containing reacting metal atoms and gaseous species on a metal support. In that case, we may write, for the sum on the chemical reacting components s in equation (2.10)

$$\sum_\gamma \nu_{\rho\gamma} a_\gamma \sigma = \sum_s \nu_{\rho s} a_s \sigma, \qquad (2.12)$$

where the partial molar surface area of γ is defined by

$$a_\gamma = \left(\frac{\partial A}{\partial n_\gamma^c}\right)_{T,p,n_{\beta\neq\gamma,\sigma}^c}. \qquad (2.13)$$

The surface pressure, which is equal to the lowering of surface tension of pure solvent (σ_1), can be defined by the heuristic relation

$$\sigma_1 - \sigma = \sum_s \frac{\alpha_s n_s^c}{A} = \sum_s \alpha_s \Gamma_s, \qquad (2.14)$$

with α_s an empirical positive constant in the considered dilute domain of composition. The coefficient α_s depends on T and on the nature of the media. For one fluctuating solute β, we thus have

$$\delta\sigma = -\alpha_\beta \delta\Gamma_\beta. \qquad (2.15)$$

By combining equations (2.13) and (2.14) we obtain

$$a_s = \frac{\alpha_s}{\sigma_1 - \sigma} \qquad (2.16)$$

and by using expressions (2.14) and (2.12), we have

$$\delta\left(\sum_s \nu_{\rho s} a_s \sigma\right) = \frac{\sigma_1}{(\sigma_1 - \sigma)^2}\sum_s \alpha_s \nu_{\rho s}\delta\sigma. \qquad (2.17)$$

Assuming a quasi-constant activity coefficient in the domain of the considered compositions, and taking into account equations (2.10), (2.17), and (2.15), we finally obtain the criterion for the chemical suface stability:

$$\tfrac{1}{2}\partial_t(\delta^2 s^c)_\rho = -T^{-1}k_\rho^c(\gamma_\gamma^c)^{m_\gamma}\cdots m_\beta(\gamma_\beta^c)^{m_\beta-2}$$

$$\times\left\{\nu_{\rho\beta}RT + \sum_s \alpha_s\nu_{\rho s}\alpha_\beta\Gamma_\beta\frac{\sigma_1}{(\sigma_1-\sigma)^2}\right\}(\delta\gamma_\beta^c)^2 \geqslant 0, \qquad (2.18)$$

whereas, in the bulk phase, equation (2.10) should be written as

$$\tfrac{1}{2}\partial_t(\delta^2 s)_\rho = -T^{-1}k_\rho\gamma_\gamma^{m_\gamma}\cdots m_\beta\gamma_\beta^{m_\beta-2}\{\nu_{\rho\beta}RT\}(\delta\gamma_\beta)^2 \geqslant 0. \qquad (2.19)$$

As can be easily seen, the sign of $\partial_t(\delta^2 s^c)_\rho$ originates either in the capillary term or in the competition between two contributions:

(1) $\nu_{\rho\beta}$, related to the autonomous character of the chemical reaction ($\nu_{\rho\beta} < 0$ for non-autocatalytic reactions and $\nu_{\rho\beta} > 0$ for autocatalytic reactions);

(2) $\sum_s \alpha_s\nu_{\rho s}$ connected to the capillary ($\nu_{\rho\beta} \gtrless 0$). If $\sum_s \alpha_s\nu_{\rho s} > 0$, the state may become unstable, even for intrinsically stable surface chemical reactions.

A detailed analysis of the criteria (2.18) has been previously shown by Sanfeld *et al.* (1990). We shall restrict our analysis here to the simple but realistic case of a condensed or extended liquid monolayer containing (or not) the solvent.

The surface dilation defined by

$$\Delta_\rho^c = \sum_\gamma \nu_{\rho\gamma}a_\gamma \qquad (2.20)$$

is almost constant. Suppose also that the activity coefficient f_γ^c is a constant quantity in the domain of the considered composition. Putting equations (2.15) into equation (2.9) one obtains finally

$$\tfrac{1}{2}\partial_t(\delta^2 s^c)_\rho = -T^{-1}k_\rho^c(\gamma_\gamma^c)^{m_\gamma}\cdots m_\beta(\gamma_\beta^c)^{m_\beta-2}\{\nu_{\rho\beta}RT + (\alpha_\beta\Gamma_\beta\Delta_\rho^c)\}(\delta\gamma_\beta^c)^2. \qquad (2.21)$$

The first term in curly brackets is the same as the classical term relative to the volume chemical stability ($\nu_{\rho\beta}RT$ in the curly brackets of equation (2.19)). An additional intrinsic surface contribution $\alpha_\beta\Gamma_\beta\Delta_\rho^c$ can, however, induce new varieties of chemical instabilities. For example, if the reaction ρ is intrinsically stable ($\nu_{\rho\beta} < 0$), the criteria for surface chemical instabilities reads

$$\left.\begin{array}{l}\Delta_\rho^c > 0, \\ \alpha_\beta\Gamma_\beta\Delta_\rho^c > RT|\nu_{\rho\beta}|,\end{array}\right\} \quad \text{instability.} \qquad (2.22)$$

If the reaction is intrinsically unstable ($\nu_{\rho\beta} > 0$), the capillary term may stabilize the chemical state and the criteria reads

$$\left.\begin{array}{l}\Delta_\rho^c < 0, \\ \alpha_\beta\Gamma_\beta|\Delta_\rho^c| > RT\nu_{\rho\beta},\end{array}\right\} \quad \text{stability.} \qquad (2.23)$$

The role played by the surfactant parameters α_β on the chemical stability is equally important, as previously shown for convective regimes by Dalle Vedove (1982). As may be seen, due to the terms σa_γ, even non-autocatalytic surface reactions are able to provoke chemical instabilities. In a certain way, the presence of the surface free energy (σa_γ) in the expression of the chemical potential characterizes the existence of a potential barrier.

Let us now illustrate the two criteria (2.22) and (2.23) by a classical experiment:

(1) Suppose a non-autocatalytic surface reaction involving quasi-infinite reservoirs of metallic compounds (Me...) reacting with only one fluctuating gas (O_2 or $X_2 \equiv S_2, \dots$). The proposed scheme is

$$Me_2X + O_2 \rightarrow 2MeO + X.$$
$$\quad O \quad X_2 \qquad \quad X \quad O$$

Equation (2.21) reads (neglecting the influence of activity coefficients)

$$\partial_t \delta^2 s^c = + \cdots \{RT - \alpha_{O_2} \Gamma_{O_2} \Delta^c\}(\delta \Gamma_{O_2})^2 \gtrless 0.$$

For a volumic reaction, one would always obtain from (2.19) a stable chemical state:

$$\partial_t \delta^2 s = + \cdots RT(\delta C_{O_2})^2 > 0.$$

On the contrary, in the monolayer, the situation can be quite different. Indeed, if $\Delta^c > 0$ and $\alpha_{O_2} \Gamma_{O_2} \Delta^c > RT$, chemical states become unstable. For example such cases occur for realistic experimental values:

$$\alpha_{O_2} = 5RT \text{ (condensed layer)},$$
$$\Delta^c \cong 3 \times 10^9 \text{ cm}^2 \text{ mol}^{-1},$$
$$\Gamma_{O_2} \cong \omega^{-1} \cong 3 \times 10^{-10} \text{ mol cm}^{-2}.$$

(2) Suppose now an autocatalytic surface reaction involving quasi-infinite reservoirs of metallic compounds (Me...) reacting with only one fluctuating gas (O or X). The proposed scheme is

$$2MeOX + O \rightarrow 2O + MeO + MeX_2$$
$$\quad X \qquad X \qquad X \qquad O_2$$

Equation (2.21) reads (neglecting the influence of the activity coefficients)

$$\partial_t \delta^2 s^c = + \cdots \{-RT - \alpha_O \Gamma_O \Delta^c\}(\delta \Gamma_O)^2 \gtrless 0.$$

For a volumic reaction, equation (2.19) gives

$$\partial_t \delta^2 s = + \cdots - RT(\delta C_O)^2 < 0.$$

The chemical state in the volume is thus always unstable, while in the monolayer a stable state might be expected if

$$\Delta^c < 0 \quad \text{and} \quad \alpha_O \Gamma_O |\Delta^c| > RT.$$

Such a situation is obtained for realistic experimental values:

$$\alpha_{O_2} = 10RT,$$
$$\Delta^c \cong -3 \times 10^8 \text{ cm}^2 \text{ mol}^{-1},$$
$$\Gamma_{O_2} \cong 5 \times 10^{-10} \text{ mol cm}^{-2}.$$

3. Shift of surface tension induced by an applied electric field

As shown in the previous sections, capillary properties play an important role on the stability of surface chemical reactions. In order to extend the same study to electro- and magnetosurface transformations (as, for example, on droplets in atmosphere) we have to study at first the influence of an external field on capillary properties. Several exhaustive contributions were published showing the condition for increasing or decreasing the surface tension far from the critical point (Rocard 1951; Defay & Sanfeld 1967; Liggieri *et al.* 1994).

More recently we extended the theory to a region near the critical point (Gouin & Sanfeld 1997). The question is more precisely to evaluate the importance of an imposed external electric field on the behaviour of a polar gas–liquid interface. Such an approach needs the use of the thermodynamic theory of polarized layers (Sanfeld 1968, p. 72). In the presence of an electric field $\boldsymbol{E}$ for polar materials,

$$\mu = \mu^* - \frac{1}{8\pi} \int_0^{E^2} \left(\frac{\partial \varepsilon}{\partial \rho}\right)_{T \cdot \boldsymbol{E}} \mathrm{d}\boldsymbol{E}^2 \quad \text{(e.s.c.g.s.u.)}, \tag{3.1}$$

where μ^* is the classical chemical potential at zero field, ε is the dielectric constant and ρ is the density. The quantity in the integral of equation (3.1) is called the polarization term. For $\partial \varepsilon / \partial \rho$ independent of ε, equation (3.1) reduces to

$$\mu = \mu^* - \frac{E^2}{8\pi} \frac{\partial \varepsilon}{\partial \rho}. \tag{3.2}$$

The extended initial conditions read

$$\frac{1}{\rho_c} \left(\frac{\partial \mu}{\partial \rho}\right)^c = \left(\frac{\partial^2 \mu}{\partial \rho^2}\right)^c = 0, \tag{3.3}$$

or

$$\frac{1}{\rho_c} \left(\frac{\partial p}{\partial \rho}\right)^c = \left(\frac{\partial^2 p}{\partial \rho^2}\right)^c = 0. \tag{3.4}$$

By a limited expansion around the critical conditions we get

$$\frac{\partial p}{\partial \rho}(\rho_{c\boldsymbol{E}}, T_{c\boldsymbol{E}}) = \rho_c \left(\frac{\partial^2 \mu^*}{\partial \rho \partial T}\right)^c (T_{c\boldsymbol{E}} - T_c). \tag{3.5}$$

The shift of the critical temperature is then deduced by

$$(T_{c\boldsymbol{E}} - T_c) = \frac{E^2}{8\pi} \rho_c \left(\frac{\partial^2 \varepsilon}{\partial \rho^2}\right)^c \Bigg/ \left(\frac{\partial^2 p}{\partial \rho \partial T}\right)^c. \tag{3.6}$$

Now in the absence of an applied electric field, σ can be expressed in terms of the spatial derivatives of ρ (Defay & Sanfeld 1974; Rowlinson & Widom 1984):

$$\sigma = \int_{\rho_g}^{\rho_l} [2c\psi(\rho, T)]^{1/2} \, \mathrm{d}\rho, \tag{3.7}$$

where ψ is the volume free-energy density. The density profile for the plane isothermal interface reads

$$C \frac{\mathrm{d}^2 \rho}{\mathrm{d}z^2} = \mu^*(\rho, T) - \mu_0^*, \tag{3.8}$$

where μ_0^* is a constant.

In the critical region Rowlinson & Widom have shown the relation

$$\mu^* = \mu_0^* - A_0(T_c - T)(\rho_c - \rho) + B_0(\rho_c - \rho)^3, \tag{3.9}$$

Table 1. *Surface tension of nitromethane near the critical point in the presence of an applied electric field*

	3×10^4 V cm^{-1}	10^5 V cm^{-1}	2×10^5 V cm^{-1}
van der Waals equation of state,	$\sigma_E = 0.02$ dyn cm^{-1}	0.7 dyn cm^{-1}	5.5 dyn cm^{-1}
CCOR polar equation,	$\sigma_E = 0.04$ dyn cm^{-1}	1.6 dyn cm^{-1}	12.6 dyn cm^{-1}

where

$$A_0 = \left(\frac{\partial^2 \mu^*}{\partial\rho\partial T}\right)^c \quad \text{and} \quad B_0 = \frac{1}{6}\left(\frac{\partial^3 \mu^*}{\partial\rho^3}\right)^c. \tag{3.10}$$

In an electric field μ^* must be replaced by μ. Considering then that the polarization term is a small quantity, which is a reasonable assumption, we then get the expression of σ in the presence of $\boldsymbol{E}$:

$$\sigma_{\boldsymbol{E}} = \frac{\sqrt{c}}{3B}[2A(T_{c\boldsymbol{E}} - T)]^{3/2}, \tag{3.11}$$

where c is a constant associated to the mean field theory ($c \approx 0.5$ erg g^{-2} cm^5).

Taking now T equal to T_c at zero field, we finally obtain

$$\sigma_{\boldsymbol{E}} = \frac{2\sqrt{c}}{(\partial^3\mu^*/\partial\rho^3)^c}\left[\boldsymbol{E}^2\frac{(\partial^2\varepsilon/\partial\rho^2)^c}{8\pi}\right]^{3/2}. \tag{3.12}$$

4. Results

In order to evaluate the shift of σ due to the field at a temperature corresponding to the critical value in the absence of the field, we use the Debye relation (Debye 1929, p. 36) and an equation of state.

The Debye relation reads

$$\frac{\varepsilon - 1}{\varepsilon + 2} = 4\pi\frac{N_{\text{Av}}\rho}{3M}\left(\alpha + \frac{\boldsymbol{m}^2}{3kT}\right), \tag{4.1}$$

where M is the molar mass, α is the polarizability and $\boldsymbol{m}$ is the permanent dipolar moment. The results summarized in table 1 for nitromethane submitted to different electric fields show important discrepancies.

Several molecules were also studied using different equations of state (van der Waals, CCOR, Bender VO) and different relations giving $\varepsilon = \varepsilon(\rho)$. As predicted by equations (3.12) and (4.1), large values of the permanent dipolar moment increase the shift of σ. In all cases a constant applied field stabilizes the gas–liquid interface. More recently, electric field effects were investigated in non-ionic fluids near the critical point (Onuki 1995). This author derives an upward shift of the critical temperature proportional to the square of the field and also an induced dipolar interaction among the critical fluctuations.

5. Conclusions

The main purpose of this investigation was to first predict the influence of capillarity on chemical stability and, secondly, the effect of an applied electric field on surface tension near the critical point. It is clearly shown by using thermodynamic criteria that a stable (unstable) reaction in a volume phase can be destabilized (stabilized) when it occurs in a monolayer spread on a liquid surface. In the second part, we evaluate the shift of surface tension due to an applied electric field by using an extended expression for the chemical potential in the frame of mean field theory. Simple equations of state combined with the Debye relation for the dielectric constant show that the field tends to stabilize the gas–liquid interface. The larger the dipole moment the larger the shift of the surface tension. Significant values of the shift are obtained for nitromethane.

Financial support has been provided by GDR 1185 CNRS/CNES and the European Union contract ERBCHRXCT 094481. The Royal Society and the Discussion Meeting organizers are thanked for the invitation that led to the writing of this paper.

References

Cesari, L. 1963 *Asymptotic behaviour and stability-problems in ordinary differential equations.* New York: Academic.

Chandrasekhar, S. 1961 *Hydrodynamic and hydromagnetic stability.* New York: Dover.

Dalle Vedove, W. & Sanfeld, A. 1982 Hydrodynamic and chemical stability of fluid–fluid reacting surfaces. *J. Colloid Interface Sci.* **3**, 1 and **85**, 43.

Debye, P. 1929 *Polar molecules.* New York: Dover.

De Donder, Th. 1942 Le concept de stabilité. *Bull Acad. R. Belg.* **41**, 946.

Defay, R. & Prigogine, I. 1966 *Surface tension and adsorption.* London: Longmans, Green.

Defay, R. & Sanfeld, A. 1967 The tensor of pressure in spherical and plane electrocapillary layers. *Electrochimica Acta* **12**, 913.

Defay, R., Sanfeld, A. & Steinchen, A. 1972 La couche interfaciale traitée comme une collection de plans moléculaires. *J. Chim. Phys.* **69**, 1380.

De Groot, S. R. & Mazur, P. 1962 *Non equilibrium thermodynamics.* Amsterdam: North-Holland.

Drazin, P. G. & Reid, W. H. 1981 *Hydrodynamic stability.* Cambridge University Press.

Duhem, P. 1911 *Traité d'énergétique*, vol. 2. Paris: Gauthier-Villars.

Gibbs, J. W. 1928 *Collected work.* New-York: Longmans, Green.

Glansdorff, P. & Prigogine, I. 1970 *Entropy, stability and structure.* New York: Wiley.

Gouin, H. & Sanfeld, A. 1997 Influence of a constant electric field on pure fluid interfaces near the critical point. *Conf. on Dynamics of Multiphase Flows across Interfaces, Wavre Meeting, Belgium, January 1997.*

Landau, L. D. & Lifshitz, E. M. 1958 *Statistical physics.* London: Pergamon.

Liggieri, L., Sanfeld, A. & Steinchen, A. 1994 Effects of magnetic and electric fields on surface tension of liquids. *Physica* A **206**, 209.

Lin, C. C. 1967 *The theory of hydrodynamic stability.* Cambridge University Press.

Minorsky, N. 1962 *Nonlinear oscillations.* Princeton, NJ: Van Nostrand.

Onuki, A. 1995 Electric-field effects in fluids near the critical point. *Europhys. Lett.* **29**, 611.

Prigogine, I. & Defay, R. 1967 *Chemical thermodynamics.* New York: Longmans, Green.

Rocard, Y. 1951 *Thermodynamique.* Paris: Masson.

Rowlinson, J. S. & Widom, B. 1984 *Molecular theory of capillarity.* Oxford: Clarendon.

Sanfeld, A. 1968 *Introduction to thermodynamics of charged and polarized layers.* London: Wiley.

Sanfeld, A. & Steinchen, A. 1971 Critères de stabilité et d'évolution des systèmes électrochimiques hors d'équilibre. *Bull. Acad. R. Belg.* **LVII**, 684.

Sanfeld, A., Passerone, A., Ricci, E. & Joud, J. J. 1990 Thermodynamic approach to competition between surface and volume reactions. *Nuovo Cim.* **12**, 303.

Sørensen, T. S. (ed.) 1979 *Interfacial instability.* Lecture Notes in Physics, vol. 105, pp. 168, 229. Berlin: Springer.

Steinchen, A. & Sanfeld, A. 1981 Thermodynamic stability of charged surfaces. In *The modern theory of capillarity* (ed. F. C. Goodrich & A. J. Rusanov), pp. 183–192. Berlin: Academic.

Discussion

A. Passerone (*ICFAM-CNR, Genoa, Italy*). Is Professor Sanfeld's theoretical approach applicable to liquid–liquid interfaces, and in particular, to systems near the clouding point?

A. Sanfeld. When dealing with two immiscible liquids the situation is rather more complicated. Indeed, even far from the critical point the two fluids interpenetrate each other in a thin layer of a about a few molecular diameters. The density of each component changes regularly between the value in both bulk phases with a profile in accordance with the one of a liquid–vapour interface. Each density vanishes asymptotically in the complementary phase. For such an interfacial layer, density gradients are important.

Phase transitions induced by electric fields in near-critical polymer solutions have been recently studied by Wirtz & Fuller (1993). They show that the mixing of two-phase solutions induced by electric fields is a universal feature shared by a wide class of systems, including upper and lower critical point polymer solutions and mixtures of low-molecular-weight molecules in a solvent. The shift of the critical temperature is quadratic in the electric-field strength.

S. K. Wilson (*Department of Mathematics, University of Strathclyde, UK*). Professor Hondros raises the important question about the control of Marangoni flows. Considerable effort has already been devoted to the influence of various additional physical effects on the onset of Marangoni convection in a quiescent layer of fluid. For example, the effects of uniform rotation of the layer (Kaddame & Lebon 1994) and a uniform field (Wilson 1993) are now reasonably well understood.

Additional references

Kaddame, A. & Lebon, G. 1994 Bénard–Marangoni convection in a rotating fluid layer with and without surface deformation. *Appl. Sci. Res.* **52**, 295–308.

Wilson, S. K. 1993 The effect of a uniform magnetic field on the onset of steady Bénard–Marangoni convection in a layer of conducting fluid. *J. Engng Math.* **27**, 161–188.

Wirtz, D. & Fuller, G. G. 1993 *Phys. Rev. Lett.* **71**, 2236.

Drops, liquid layers and the Marangoni effect

Manuel G. Velarde

Instituto Pluridisciplinar, Universidad Complutense de Madrid,
Paseo Juan XXIII, n.1, 28040-Madrid, Spain

An overview is given of recent results about the onset and development of steady and time-dependent flow motions past an instability threshold induced by the Marangoni effect. First, I consider the case of a liquid drop immersed in another immiscible liquid when (endo- or exothermic) reaction, heat and/or mass transfer at/across the drop surface, etc., leads to self-propelled drop motion, overcoming viscous drag. Then I recall salient features about the spreading of an immiscible or a dissolving drop (with surfactant) on the surface of another liquid. Finally, I consider Bénard layers when either steady convective patterns or waves are produced by appropriate heat or mass transfer across the open surface.

Keywords: interfacial instability; spreading; Bénard patterns; interfacial waves; dissipative solitons; drop motions

1. Introduction

When there is an open surface or an interface exists between two liquids, the interfacial tension, σ, accounts for the jump in normal stresses proportional to the surface curvature across the interface; hence this Laplace force affects its shape and stability. Gravity competes with it in accommodating equipotential levels with curvature. Their balance defines, for instance, the stable equilibrium of spherical drops or bubbles. If the surface tension varies with temperature or composition, and, eventually, with position along an interface, its change takes care of a jump in the tangential stresses. Hence its gradient acts like a shear stress applied by the interface on the adjoining bulk liquid (Marangoni stress), and thereby generates flow or alters an existing one (Marangoni effect). Surface tension gradient-driven flows are known to affect the evolution of growing fronts, and measurements of transport phenomena. The variation of surface tension along an interface may be due to the existence of a thermal gradient along the interface or perpendicular to it. In the former case we have instantaneous convection while in the latter flow occurs past an instability threshold (Levich 1965; Levich & Krylov 1969; Scriven & Sternling 1960; Ostrach 1982; Davis 1987). The Marangoni effect is the engine transforming physicochemical energy into flow, whose form and time dependence for standard liquids rests on the sign of the thermal gradient and the ratios of viscosities and diffusivities of adjacent fluids.

2. Drops and bubbles

(a) 'Passive' drops

To place in context our recent findings and the difficulties still existing ahead of us, let me recall how the hydrodynamic force on a drop has been estimated since the pioneering work of Newton and Stokes (Levich 1965; Levich & Krylov 1969; Happel & Brenner 1965; Edwards *et al.* 1991). From Newton's experiments in 1710, and later observations, the magnitude of the drag force of a viscous fluid on a solid sphere (a drop, in a first approximation) in steady motion was given as

$$F_D = 0.22\pi R^2 \rho U^2, \tag{2.1}$$

where U is the relative velocity between particle and fluid, R is the particle radius and ρ is the fluid density. This relation is for 'large' values of U, for which inertial (kinetic theory) effects are dominant.

Stokes, in 1850, suggested that at very low relative velocities all inertial effects are so small that they can be omitted from the Navier–Stokes equations (creeping flow approximation). Under this condition, the drag on a sphere is

$$F_D = 6\pi R\eta U, \tag{2.2}$$

with μ denoting the dynamic or shear viscosity of the fluid. Oseen pointed out that at a great distance from the sphere the inertia terms become more important than the viscous terms, and suggested a possible improvement of the Stokes's law (2.2) by taking inertial terms partly into consideration. He obtained a drag force,

$$F_D = 6\pi R\eta U(1 + \tfrac{3}{16}Re), \tag{2.3}$$

where ν is the kinematic viscosity of the fluid and $Re = (2R)U/\nu$. Neither Stokes's nor Oseen's laws are *uniformly* valid, and the latter is not really an improvement of the former. Rather Stokes's analysis is valid in a small enough neighbourhood of the sphere and Oseen's analysis, though valid far from the sphere, is not valid when approaching the sphere. Oseen's approximation, although incorporating inertial terms, is a linear theory and Stokes's approximation is a steady-state theory.

The mathematical problems solved by Stokes and Oseen come from different approximations to the Navier–Stokes equations, together with appropriate initial and boundary conditions. As for the axisymmetric motion of a sphere there is always a stream function, Ψ, with (r, θ, ϕ) coordinates and the hydrodynamic equations reduce to (Levich 1965; Happel & Brenner 1965)

$$\frac{\partial}{\partial t}(E^2\Psi) + \frac{1}{r^2\sin\theta}\frac{\partial(\Psi, E^2\Psi)}{\partial(r,\theta)} - 2\frac{E^2\Psi}{r^2\sin^2\theta}\left(\frac{\partial\Psi}{\partial r}\cos\theta - \frac{1}{r}\frac{\partial\Psi}{\partial\theta}\sin\theta\right) = \nu E^4\Psi, \tag{2.4}$$

with

$$\frac{\partial(f,g)}{\partial(r,\theta)} = \begin{pmatrix} \dfrac{\partial f}{\partial r} \dfrac{\partial f}{\partial\theta} \\ \dfrac{\partial g}{\partial r} \dfrac{\partial g}{\partial\theta} \end{pmatrix}$$

and

$$E^2 = \frac{\partial^2}{\partial r^2} + \frac{\sin\theta}{r^2}\frac{\partial}{\partial\theta}\left(\frac{1}{\sin\theta}\frac{\partial}{\partial\theta}\right).$$

The Stokes aproximation is the reduction of equation (2.5) to

$$E^4\Psi = 0. \tag{2.5}$$

Equation (2.6) with the appropriate non-penetration no-slip/stick BC and suitable asymptotic behaviour for large r, yields

$$\Psi(r,\theta) = (-\tfrac{1}{4}U(R^3/r) + \tfrac{3}{4}URr - \tfrac{1}{2}Ur^2)\sin^2\theta, \tag{2.6}$$

hence the hydrodynamic drag force (2.2). Note that from (2.6) it follows that the disturbance of the sphere extends to infinity as $1/r$, and the presence of a boundary or another drop can modify the flow appreciably even when placed at a distance of many diameters from the drop. It was not until the fifties that Stokes's and Oseen's results were properly put in context and generalized.

In 1911–12, Rybczynski and Hadamard, independently, solved the Stokes problem for a liquid drop with flows outside and inside (Levich 1965; Happel & Brenner 1965). Their extension of equation (2.2) is

$$F_{\mathrm{D}} = 6\pi\eta_0 U R \left[1 + \frac{\frac{2}{3}\mu}{1+\mu}\right], \tag{2.7}$$

with $\mu = \eta_{\mathrm{i}}/\eta_{\mathrm{o}}$ ($\eta_{\mathrm{i}} = \eta_{\mathrm{drop}}$). The limit η_{i} going to infinity yields Stokes's law (2.2), while $\eta_{\mathrm{o}} \gg \eta_{\mathrm{i}}$ yields the corresponding law for a bubble, with 4 rather than 6 in equation (2.2).

In 1957, Proudman & Pearson (1957) considered Stokes's solution as a local (or inner) solution of the problem and Oseen's as the regular (or outer) solution. The former was assumed to be valid in a spherical region of radius $1/Re$ around the sphere while the outer solution was valid from infinity down to the $1/Re$ neighbourhood. In the overlapping zone both solutions were accepted as valid; hence the need to appropriately match them. Proudman & Pearson found that for non-vanishing albeit low Reynolds number flows ($Re \ll 1$) the hydrodynamic drag on the sphere is

$$F_{\mathrm{D}} = 6\pi\eta U R(1 + \tfrac{3}{16}Re + Re^2 + \tfrac{9}{160}Re^2 \ln Re + \cdots), \tag{2.8}$$

which shows the non-analytic form of the expansion.

Subramanian (1981) used matched asymptotic expansions to obtain the hydrodynamic force on a drop including convective terms in the heat equation while maintaining the Stokes approximation for the velocity field. With

$$M = -\left(\frac{\partial\sigma}{\partial T}\right)\frac{\mathrm{d}\delta T}{\kappa\eta},$$

where κ is the thermal diffusivity of the drop, T denotes temperature and ΔT is the temperature contrast over a length scale d (this scale may be taken as R), his series expansion in terms of the Marangoni (as a Péclet, $Pe = 2RU_{\mathrm{T}}/\kappa$, $U_{\mathrm{T}} = -(\partial\sigma/\partial T)\Delta T/\eta$) number, and subsequent improvement by his collaborator Merritt, did not show any logarithmic term (Merritt & Subramanian 1988; Subramanian 1992; Wozniak et 1988). This was due to the way the outer solution was treated.

Taylor & Acrivos (1964) obtained the contribution of the deformation of the sphere in terms of the capillary number. At $Re = 0$, a drop or a bubble remains spherical irrespective of the low or high value of the (constant) surface tension. However, deformation may be relevant even when inertial effects are ignored if the surface tension σ is not constant. If over the spherical surface $\Delta\sigma$ is small compared to the value of σ, the capillary number will be small, and the drop or bubble may be asumzed spherical with negligible error.

Young *et al.* (1959) were the first to realize the possibility of levitating a drop or a bubble by Marangoni stress. They showed that a drop or a bubble placed in another fluid where a temperature gradient exists instantaneously tends to move towards the hotter point. This is the motion of the drop relative to the flow induced along its surface by the lowering of surface tension at its leading pole (hotter than the rear pole). Using the Stokes–Rybczynski–Hadamard approximation they also computed the terminal velocity of a drop or a bubble in the field of gravity, and experimentally checked the theoretical prediction within reasonable accuracy (within 20%) with an experiment using rising bubbles in a liquid layer heated from below (diameters $2R = 10^{-3}$–22×10^{-3} cm; $\nabla T = dT/dz = 10$–90 K cm^{-1}; z denotes the vertical coordinate). Using neutrally buoyant liquid water at $4\,^\circ$C, Bratukhin *et al.* (1979, 1982) did a similar experiment with rising bubbles in a laterally heated vertical liquid layer.

In their experiment, Young *et al.* (1959) used a tiny open liquid bridge. An improvement eliminating wetting and capillary convection at the open sides was carried out by Hardy (1979). He used a closed cavity with silicone oil and air bubbles ($2R = 5$–25×10^{-3} cm; $dT/dz = 40$–140 K cm^{-1}). Hardy discussed the role of contamination at the surface of the bubble, earlier suggested by Levich (1965; Levich & Krylov 1969). Further improvement came with an experiment by Merritt & Subramanian (1988). Experimentalists started using drops rather than bubbles. Barton & Subramanian (1989) used neutrally buoyant drops ($2R = 20$–600 mm, $dT/dz = 2.4$ K mm^{-1}). Recent Earth-based and low-g work by Braun *et al.* (1994) on thermocapillary migration of drops provided the most accurate verification of the prediction by Young *et al.* (1959). Their Marangoni (Péclet) number was in the range 10^{-5}–10^{-6}, but with a a non-standard liquid of surface tension increasing with the increase of temperature (2 butoxyethanol-water mixture with liquid–liquid phase separation at $61.14\,^\circ$C on the lower branch of the closed miscibility gap; $2R = 11$ μm, $dT/dz = 36.9$ K m^{-1}, $d\sigma/dT > 0$).

For the Young, Goldstein and Block problem the balance between capillary, buoyancy and hydrodynamic forces is

$$F_\sigma + F_g = AU, \tag{2.9}$$

with the (Marangoni) capillary force

$$F_\sigma = \frac{4\pi R^2}{(1+\mu)(1+\delta)}\left(\frac{d\sigma}{dT}\right)(\nabla T)_\infty \tag{2.10}$$

and the buoyancy force

$$F_g = \tfrac{4}{3}\pi R^3 g(\rho_i - \rho_o). \tag{2.11}$$

As

$$A = 4\pi\eta_0 \frac{1+\tfrac{3}{2}\mu}{1+\mu}R > 0, \tag{2.12}$$

the hydrodynamic force AU represents drag. $\delta = \lambda_i/\lambda_o$ is the ratio of thermal conductivities (drop to surrounding fluid). Clearly, AU embraces both Stokes's law ($\eta_i \to \infty$, $\mu \to \infty$) and the Rybzcynski–Hadamard law ($[\nabla_T]_\infty = 0$).

(b) 'Active' drops

The work on drops and bubbles so far recalled, refers to 'passive' drops. Now I turn to some recent findings about 'active' drops obtained with A. Ye. Rednikov and Yu. S. Ryazantsev (Rednikov *et al.* 1994a–e, 1995a; Velarde *et al.* 1996). By an 'active' drop or bubble I mean a drop or a bubble with internal volume heat sources, with a surface where chemical reactions may occur, or there is drop dissolution with heat release, etc. Take, for example, a drop at rest in a homogeneous medium and assume that there is (uniform) internal heat generation or a surface chemical reaction. Let us evaluate how the state of rest can be made unstable. Consider, for instance, the latter case with given uniform composition far off the drop. A composition fluctuation at the surface of the drop brings the Marangoni effect which yields flows inside and outside the drop. They can be sustained if the Marangoni effet is strong enough relative to viscous drag and heat diffusion, i.e. past an instability threshold. Indeed, as the drop moves the flow brings to the leading pole the higher solute at the surface, it makes the concentration far off the drop higher than that in its vicinity. Alternatively, if a velocity fluctuation tending to move the drop in a certain direction spontaneously occurs, it breaks the initial spherical symmetry in composition; hence the possibility of bringing strong enough Marangoni stresses which in turn can help sustain the velocity fluctuation past an instability threshold. If the initial state is that of a uniform constant drop velocity or there is an externally imposed temperature or composition gradient, as in the experiment of Young *et al.* (1959), then instability is also possible leading to a different drop motion. These are not the only possible instabilities as the effects due to deformability and time dependence may come into play.

To illustrate how we have proceeded along the path set by the earlier mentioned authors dealing with passive drops, now I consider the case of a spherical drop moving with constant velocity in a temperature and composition-homogeneous infinitely extended surrounding fluid. Both the inner and outer fluids are taken immiscible. The surface tension is assumed to vary linearly with temperature. The outer fluid is assumed to have a uniform concentration of a solute which is allowed to react exo- or endothermally at the surface of the drop. Far off the drop both the temperature and concentration of the solute are constants; hence there is no external gradient. Stefan flow is negligible (convective flow of the reacting components in a direction normal to the surface where the reaction is taking place; it is generally a small effect for most chemical reactions and is normally important only in the presence of strong ablation or condensation). Let us consider the low Reynolds and Péclet number approximations with, however, $M Pe = 1$ (M is defined as earlier for the drop while Pe is defined using the far-field reference velocity, U, as for the earlier used Reynolds number, with indeed κ replacing ν). In dimensionless form the steady equation (2.4) becomes

$$\frac{Re}{\nu^* r^2} \left\{ \frac{\partial(\psi_i, E^2 \psi_i)}{\partial(r, \delta)} + 2E^2 L_r \psi_i \right\} = E^4 \psi_i, \tag{2.13}$$

which is nonlinear. It is considered together with the corresponding nonlinear heat and mass diffusion equations, and appropriate BC. Here Re is defined with the constant drop velocity, $\delta = \cos\theta$, ν^* refers to (kinematic) viscosity ($i = 1$, outer fluid, $\nu^* = 1$; $i = 2$, drop, $\nu^* = \nu$), and L_r corresponds to the operator appearing in the third term of equation (2.4).

The linear solution of equation13) yields the hydrodynamic force (Rednikov *et al.* 1994*a*–*c*; Velarde *et al.* 1996).

$$F = -4\pi \eta_1 r A U. \tag{2.14}$$

If A is negative we have drag while if A is positive there is thrust; hence self-propulsion and autonomous motion of the drop in a medium originally uniform. We have

$$A = -[1 + \tfrac{3}{2}\mu + 3m]/(1 + \mu + m), \tag{2.15}$$

with $\mu = \eta_1/\eta_2$ and m a suitably scaled Marangoni number accounting for a balance between $d\sigma/dT$, the chemical reaction rate, and (viscous and heat) dissipation. For further details and the various cases studied see Rednikov *et al.* (1994a–e, 1995a) and Velarde *et al.* (1996). We see that at $m = m_1 = -\frac{1}{3} - \frac{1}{2}\mu$ we have $A = 0$, where, as at $m = m_2 = -1 - \mu$, A diverges to infinity. Clearly, the study of both cases demands nonlinear analyses. Before referring to this we note that for $A = 3$, the Marangoni effect combined with the chemical reaction yields the possibility of self-propulsive thrust, while for $A = -1$ we have drag, as well as for $A = -\frac{3}{2}$. In the later case, the flow inside the drop completely stagnates as if it were a solid sphere. For $A = -3$ there is enhanced drag due to the appearance of recirculating backflow around the drop and flow reversed in the drop.

The weakly nonlinear result for $(m - m_1)$ $(> 0, \ll 1)$ provides the value of the hydrodynamic force

$$F = -4\pi\eta_1 R[A + B(RU/\nu_1)]U, \tag{2.16}$$

with

$$B = -\tfrac{1}{4}A^2 + \tfrac{1}{2}Pr[m(6 + 3L - AL)/(1 + \mu + m)], \tag{2.17}$$

where here $Pr = \nu_1/\kappa_1$, and $L = \kappa_1/D$ is the (inverse) Lewis number of the homogeneous surrounding medium. A representation of F versus U straightforwardly shows that for $(m - m_1) < 0$ $(A > 0)$ there are three possible values of U for zero hydrodynamic force; hence, possible autonomous motion of the drop. The addition of an external force field like buoyancy, or an externally imposed thermal gradient like in the experiment by Young *et al.* (1959) provides the possibility of three genuinely different non-zero velocities for a given force field strength. We have multiplicity of steady states of motion which cannot all actually be realizable.

For $m < m_2$, there are three possible values of the hydrodynamic force for zero velocity; hence three possible coexisting levitation levels of which one cannot be stable. Levitation or motion is a consequence of the nonlinearity in the flow coupled to the Marangoni stress, and not a consequence of some external thermal gradient as in the case treated by Young *et al.* (1959).

Complete stability analyses, respectively, for m around m_1 and m around m_2, have not been carried out. They demand the inclusion of the time-dependent term in equation (2.4) and explicit use of the convective state, which is a formidable task. However, ad hoc quasi-stability analyses have confirmed the results reported here. For further details see Rednikov *et al.* (1994a–c).

Finally, addition of a time-varying gravity field as it occurs in space (*g*-jitter) and in some Earth-based experiments, leads to a time-dependent weakly nonlinear vector-form Landau equation for the velocity of the drop (Rednikov *et al.* 1995). This equation can be used not only to find the stationary regimes and analyse their stability, but also to consider time-varying motions. In the simplified case of a small amplitude buoyancy force changing sinusoidally with time the result found is that an *active* drop capable of *autonomous* motion actually tends to move in a direction orthogonal to the time-varying force.

3. Spreading of a drop on another liquid under Marangoni stress

(a) Insoluble surfactant

With de Ryck and Starov (Starov *et al.* 1997), I have recently considered the spreading of an insoluble drop of surfactant over a thin viscous layer (e.g. less than a millimeter depth). The experiment and theory refer to a case where diffusion is a much slower process than convection due to the Marangoni effect. For concentrations above CMC (about $3\,\mu l$ of an aqueous solution of SDS at $c = 20\,g\,l^{-1}$; CMC is $4\,g\,l^{-1}$) the spreading process involves two stages. First, there is a rather fast stage when the surfactant concentration is determined by the dissolution of micelles, with a duration fixed by the initial amount of micelles. Then it follows a second slower stage when the surfactant concentration drastically changes in the original center position of the drop but as the total mass of surfactant remains constant a hole, a dry spot is created in the liquid layer.

As the drop circularly spreads due to the dominant role played by the nonlinearity, the percursor film develops a shock-like front whose time evolution according to theory follows a power law in time; first about $\frac{1}{2}$ (experimentally, $t^{0.6}$) and subsequently $\frac{1}{4}$ (experimentally, $t^{0.17}$). The radius of the inner hole at the centre expands, also with a power law $t^{1/4}$ (theory and experiment). Further theoretical developments for the more general case when diffusion takes on a time scale comparable to Marangoni convection can be found in Starov *et al.* (1997), Borgas & Grotberg (1988) and Gaver & Grotberg (1990, 1992).

(b) Soluble surfactant

Another interesting experiment with, however, a dissolving drop of nitroethane deposited over a liquid water layer has been recently conducted by Santiago-Rosanne *et al.* (1997). Nitroethane has a much lower surface tension ($39\,mN\,m^{-1}$) than water being only partially miscible in it in proportions lower than 5% at room temperature. Here too the Marangoni effect plays a dramatic role in creating shock-like fronts as well as smooth solitary-like waves. Indeed, the deposition of the nitroethane drop induces a sudden local change of surface tension at the circular periphery; hence the dramatic outward front motion. The interplay of Marangoni stresses and gravity creates daisy-like patterns which are in fact time-dependent structures. The daisy flower petals result from collisions of wave crests. Wave profiles and the kinematics of collisions of both the smooth solitary-like waves and the shocks have been studied and qualitative agreement exists with the theory I have recently developed, leading to dissipative Marangoni-driven Boussinesq–Burgers–Korteweg-de Vries (KdV) equations. Further below I return to this problem when discussing wave motions in Bénard layers. Details of the experiments with good diagrams and theory can be found in Santiago-Rosanne *et al.* (1997).

4. Bénard layers: convective patterns

The onset of patterned convective motions in heated fluid layers with a free upper surface has been extensively studied since the original experiments by Bénard (Koschmieder 1993; Velarde & Normand 1980; Normand *et al.* 1977). Depending on the depth of the layer, d, one distinguishes two basic mechanisms of instability. In sufficiently deep layers or in containers where the fluid is confined between rigid horizontal plates, the convective motion settles when buoyancy forces overcome viscous forces and heat dissipation (Rayleigh–Bénard problem) (Normand *et al.* 1977). Alternatively, in shallow enough layers with an open surface, inhomogeneity in the surface tension, hence the Marangoni effect is responsible for the onset of motion (Bénard–Marangoni problem). In both cases, the characteristic wavelength of the convective structure is about the depth of the cell or much larger, depending on whether or not the horizontal boundaries are good thermal conductors. Close to the instability threshold the system may be described by amplitude equations whose coefficients depend on the dimensionless numbers of the problem containing fluid properties, boundary conditions and the external forcing. Most of these numbers have been earlier defined. If the Biot number, $Bi = hd/\kappa$, where h is the thermal surface conductance, its infinite value corresponds to a perfectly conducting boundary while a zero value corresponds to a poorly conducting surface.

To the Navier–Stokes, continuity and energy equations, in the Boussinesq approximation (Normand *et al.* 1977), we add the boundary conditions. At the lower uniformly heated rigid plate, $v = 0$ and $\partial T/\partial z = Bi\,T$. At the top open surface, $w = 0$, $\partial\sigma/\partial z = \eta(\partial u/\partial z)$, $\partial\sigma/\partial y = \eta(\partial v/\partial z)$ and $\partial T/\partial z = -\,Bi\,T$, where w is the vertical velocity component.

To study the transition between the motionless state and convection, and the dynamics of the structures that define this convective state, a multiple scale perturbation theory has been developed in the vicinity of the onset of convection (Bragard *et al.* 1996; Bragard & Velarde 1997, 1998, where reference is also given of related recent work by other authors). A small parameter allows separation of the fast variables that describe the instability and the slow variables that describe the pattern dynamics. For instance, the temperature can be written as

$$T = T(z)[A_1(X,Y,\tau)\exp(ik^{(1)}\cdot r) + A_2\exp(ik^{(2)}\cdot r) + A_3\exp(ik^{(3)}\cdot r) + \text{c.c.}], \quad (4.1)$$

where $k^{(i)}$ denotes three linearly critical wave vectors oriented at $120°$ in the horizontal plane. The amplitude equations in the horizontal plane are (e.g. for A_1)

$$\begin{aligned}
\alpha_t\partial_t A_1 = {} & \alpha_t\Delta A_1 + \alpha_q A_2^* A_3^* \\
& -\alpha_{cs}A_1|A_1|^2 - \alpha_{ci}A_1(|A_2|^2 + |A_3|^2) \\
& +\alpha_d(k^{(1)}\cdot\nabla_x)^2 A_1 + i\beta_1(K^{(1)}\cdot\nabla_x)(A_2^* A_3^*) \\
& +i\beta_2[A_2^*(k^{(2)}\cdot\nabla_x)A_3^* + A_3^*(k^{(3)}\cdot\nabla_x)A_2^*] \\
& +i\beta_3[A_3^*(k^{(2)}\cdot\nabla_x)A_2^* + A_2^*(k^{(3)}\cdot\nabla_x)A_3^*], \quad (4.2)
\end{aligned}$$

where $\alpha_1 = 0.0038$, $\alpha_t = 0.05 + 0.013\,Pr^{-1}$, $\alpha_q = 0.0203 - 0.0046\,Pr^{-1}$, $\alpha_{cs} = 0.016 + 0.0049\,Pr^{-1} + 0.00077\,Pr^{-2}$, $\alpha_{ci} = 0.0217 + 0.003\,Pr^{-1} + 0.0018\,Pr^{-2}$, $\alpha_d = 0.0021$, $\beta_1 = \beta_2 = \beta_3 = \beta = 0.0016 - 0.0041\,Pr^{-1}$ and $\Delta = M - M_c$. Similar equations appear for A_2 and A_3 (with circular permutation of the indices) (Bragard & Velarde 1997, 1998). The numbers correspond to the specific case of a poor conducting upper surface and good conducting lower plate as in standard experiments. These equations are generalized Ginzburg–Landau equations with advective terms with nonvanishing β coefficients. In general, for these equations there is no Lyapunov functional; hence for some value of the β we may observe no steady behaviour. Δ and β measure the (sub/supercritical) distance to the threshold and the strength of the advective terms, respectively. In the simpler case of $Pr \to \infty$ we define

$$\Delta_c = -\frac{\alpha_q^2}{4\alpha_1(\alpha_{cs} + 2\alpha_{ci})} \approx -0.456, \quad (4.3)$$

$$\Delta_1 = -\frac{\alpha_q^2\alpha_{cs}}{\alpha_1(\alpha_{cs} - \alpha_{ci})^2} \approx -53.4, \quad (4.4)$$

$$\Delta_2 = -\frac{\alpha_q^2(2\alpha_{cs} + \alpha_{ci})}{\alpha_1(\alpha_{cs} - \alpha_{ci})^2} \approx 179.2. \quad (4.5)$$

Computations and stability analysis show two hysteresis cycles; hence coexistence and bistability appears in the intervals $[\Delta_c, 0]$ and $[\Delta_1, \Delta_2]$, respectively.

As an illustration, for a square container starting from random initial conditions (cases (i)–(iv)) or rolls (v) we have the following results.

(i) ($\Delta = 50$, $\beta = 0.1$). The system evolves to a stationary hexagonal pattern. Relative to the case $\beta = 0$ an increase of β only slightly distorts the pattern. The fluid rises in the centre of the cells in accordance with experimental observations. No defects are observed.

(ii) ($\Delta = 75$, $\beta = 0$). The system is in a bistable region of hexagons and rolls.

(iii) ($\Delta = 150$, $\beta = 0$). The rolls are the preferred structure. At boundaries the rolls tend to be perpendicular to the sidewalls.

(iv) ($\Delta = 150$, $\beta = 0.1$). The system does not reach a steady state. Besides the roll structure, defects appear moving through the system.

(v) ($\Delta = 150$, $\beta = 0.1$). Here the computation is with the same parameter values as in (iv) but with rolls plus noise added as an initial condition. First, the structure evolves to rolls without defects, but as time goes on the rolls start to bend leading to a 'zig-zag'-like instability (Manneville 1990). The system does not show evolution to a steady state, but rather tends to a labyrinthine structure and possibly interfacial 'turbulence'.

5. Bénard layers: overstability and waves

Let us consider now the liquid layer heated from the air side or open to suitable mass adsorption from a vapor phase above, with subsequent absorption in the bulk; hence creating a (stabilizing) thermal gradient inside the liquid layer. Contrary to the case of a layer heated from the liquid side, here the layer is stably stratified and the problem refers to oscillatory motions, waves, and not to Bénard cells (Levchenko & Chernyakov 1981; García-Ybarra & Velarde 1987; Chu & Velarde 1988).

(a) Oscillatory flow motions (transverse and longitudinal waves, surface and internal waves and their mode mixing)

Generally, the Bénard problem with Marangoni stresses, gravity and buoyancy involves several time scales. On one hand we have the viscous and thermal scales, $t_{\mathrm{vis}} = d^2/\nu$, $t_{\mathrm{th}} = d^2/\kappa$, respectively. There also exist two time scales associated with gravity and surface tension (Laplace force) that tend to suppress surface deformation, $t_{\mathrm{gr}} = (d/g)^{1/2}$, $t_{\mathrm{cap}} = (\rho d^3/\sigma)^{1/2}$ (g is gravity acceleration and ρ is density or density contrast; other quantities as earlier defined). The time scale related to the Marangoni effect is $t_{\mathrm{Mar}} = (\rho d^2/|\sigma_{\mathrm{T}}\beta|)^{1/2}$ ($\sigma_{\mathrm{T}} = \mathrm{d}\sigma/\mathrm{d}T < 0$, $\beta = \Delta T/d$, $\beta > 0$ when heating from below). There is also another time scale related to buoyancy due to the stratification imposed by the temperature gradient, $t_{\mathrm{st}} = (1/(|\alpha\beta|g))^{1/2}$ (α is the thermal expansion coefficient, positive in the standard case). The ratios of the time scales give rise to the dimensionless groups

$$Pr = \frac{t_{\mathrm{th}}}{t_{\mathrm{vis}}} = \frac{\nu}{\kappa}, \quad M = \sigma_{\mathrm{T}}\frac{t_{\mathrm{th}}t_{\mathrm{vis}}}{t_{\mathrm{st}}^2} = -\frac{\sigma_{\mathrm{T}}\beta d^2}{\eta\kappa}, \quad R = \frac{t_{\mathrm{th}}t_{\mathrm{vis}}}{t_{\mathrm{st}}^2} = \frac{\alpha\beta g d^1}{\nu\kappa},$$

$$G = \frac{t_{\mathrm{th}}t_{\mathrm{vis}}}{t_{\mathrm{gr}}^2} = \frac{gd^3}{\nu\kappa}, \quad B = \frac{t_{\mathrm{cap}}^2}{t_{\mathrm{gr}}^2} = \frac{\rho g d^2}{\sigma},$$

which heuristically are the earlier introduced Prandtl, Marangoni, Rayleigh, Galileo and (static) Bond numbers, respectively.

These time scales are not always of the same order (accordingly, Pr, M, R, G and

B are not always of order unity). For example, for the Pearson problem (Pearson 1958), i.e. the onset of the monotonic instability in a liquid layer with undeformable surface, we have $M \approx 1$, but $G \gg 1$ for gravity to be able to keep the surface practically level, whatever flows and thermal inhomogeneities exist, and $R \ll M$ for buoyancy to be neglected. The characteristic time scale of the problem is essentially $t_{\text{th}} \approx t_{\text{vis}} \approx t_{\text{Mar}}$ (at $Pr \approx 1$).

Thus, for monotonic instability there exists a finite limit of the critical Marangoni number as $G \to \infty$ given just by Pearson's (1958) results. However, this is not the case for the oscillatory instability waves, where the critical Marangoni number tends to infinity with $G \to \infty$. Indeed, oscillatory instability does not appear in the one-layer problem with undeformable surface. Thus the critical Marangoni number should better be scaled with G, as G becomes very large ($G \to \infty$).

It is also known that for high enough values of G an oscillatory mode is the capillary-gravity wave. The time scales t_{gr} and t_{cap} associated with this wave are much smaller than the viscous and thermal time scales (at least for $Pr \approx 1$, $B \approx 1$). Then dissipative effects are weak and hence the wave is very much the inviscid liquid capillary-gravity wave with the dispersion relation $\omega^2 = G \, Pr \, k(1 + (k^2/B)) \tanh(k)$ (to non-dimensionalize ω the thermal time scale is used hereafter; k is the dimensionless wave number in units of d^{-1}). Then the problem of oscillatory instability partly reduces to the question whether the Marangoni effect can sustain this otherwise damped wave. Clearly, the higher the G (and the wave frequency), the stronger should be the work of the Marangoni stresses (i.e. the critical Marangoni number) to sustain it, in agreement with the argument of the previous paragraph. An oscillatory instability can indeed be associated with the capillary-gravity wave. For a standard liquid layer, σ_{T}, this instability appears when heating the liquid layer from the air side ($M < 0$), as expected.

However, when the Marangoni number is high enough (and negative), there exists another high-frequency oscillatory mode. Indeed when a liquid element rises to the surface, it creates a cold spot there. Then, the surface tension gradient acts towards this spot, pushing the element back to the bulk, leading to overstability. High values of M ensure that the oscillations exist. Their characteristic time scale is t_{Mar}. The corresponding wave is called 'longitudinal' as it is due to the Marangoni stresses along the surface in contrast to capillary-gravity waves with essentially transverse motion of the surface. Lucassen made the theoretical (and experimental) discovery of this wave mode (Lucassen 1968).

Calculation yields the following expression for the frequency of the longitudinal wave (in the limit $M \to -\infty$):

$$\omega^2 = -M \frac{Pr}{Pr^{1/2} + 1} k^2.$$

Although this longitudinal wave has a genuinely dissipative nature, the damping rate proves to be asymptotically smaller, $O(|M|^{1/4})$, than its frequency. In practical terms the flow field accompanying the longitudinal wave is qualitatively similar to that for the capillary-gravity wave. Potential flow can be assumed in the bulk of the layer, while vorticity is present only in boundary layers at the bottom rigid plate and at the upper free surface. The boundary layer thickness is of the order $O(|M|^{-1/4})[O(G^{-1/4})]$ for the capillary-gravity wave. However, there is a significant difference, as already noted by Lucassen (1968). For the longitudinal wave the horizontal velocity field in the surface boundary layer is much more intense than the potential flow in the bulk (by $O(|M|^{1/4})$) at variance with the capillary-gravity wave. Thus, the longitudinal motion is really concentrated near the surface. Furthermore, it appears that with an undeformable surface ($1 \ll |M| \ll G$), the longitudinal mode is always damped. Indeed, oscillatory instability does not appear in the one-layer Marangoni problem without surface deformability. However, if the longitudinal wave is accompained by non-negligible surface deformation ($|M| \geqslant G$), it can be amplified, a striking result (Rednikov *et al.* 1998*a*).

Thus, at $G \gg 1$, the oscillatory Marangoni instability is associated with two high-frequency wave modes; capillary-gravity and longitudinal. The damping rate of one wave mode cannot be drastically changed, or even converted into amplification (oscillatory instability), if the underlying framework for the other wave mode does not operate. As already stated, to sustain the longitudinal wave one needs surface deformability. It also occurs that to sustain a capillary-gravity wave one needs the Marangoni effect. This is the coupling between capillary-gravity and longitudinal waves. The latter necessarily implies viscous effects. Presumably, the most dramatic manifestation of this coupling occurs at resonance, when the frequencies (2.1) and (2.2) are equal to each other. Near resonance there is mode-mixing. Namely, the capillary-gravity mode in the parameter half-space from one side of the resonance manifold is swiftly converted into the longitudinal mode when crossing the manifold, and vice versa. Another feature of resonance is that the damping/amplification rates are drastically enhanced here ($O(G^{3/5})$ versus $O(G^{1/4})$ far from resonance) (Rednikov *et al.* 1998*a*).

If the liquid layer is deep enough and has an undeformable surface the possibility also exists of coupling longitudinal waves to internal (negative buoyancy-driven) waves with $|R| \ll G$. This is the Rayleigh–Marangoni problem. Indeed, the role of the capillary-gravity wave is now played by the Brunt–Väisälä internal wave of frequency

$$\omega^2 = -R\,Pr\,\frac{k^2}{k^2 + \pi^2 n^2} \quad (n = 1, 2, \ldots),$$

whose existence is due to the stable stratification induced by heating the layer from above. The oscillatory instability comes again from wave–wave coupling, now between internal and longitudinal waves. Note that in the absence of the Marangoni effect, no oscillatory motion via instability is possible, which again stresses the crucial role played by the coupling. This instability is studied in detail in Rednikov *et al.* (1998*b*).

Although general features of mode coupling are similar in the two problems, there are differences. In the Rayleigh–Marangoni case we have a countable number ($n = 1, 2, \ldots$) of internal wave modes, and the longitudinal wave can be coupled to each of them; hence there is a countable number of marginal stability conditions. The form of the marginal curves is qualitatively different. Furthermore, there exists the minimally possible Rayleigh number (in absolute value), below which there is no oscillatory instability. No such bound was found for the Galileo number in the first problem (at least in the region where G remains high).

(*b*) *Nonlinear waves*

(i) *Nonlinear theory for long wavelength motions in shallow layers*

The nonlinear evolution past threshold of either capillary-gravity or longitudinal waves poses formidable tasks. Let us then concentrate on a simplified analysis, which is still amenable to experimental test. If we restrict consideration to just long wave oscillatory instability it has been shown (Chu & Velarde 1991; Velarde *et al.* 1991; Garazo & Velarde 1991; Nepomnyashchy & Velarde 1994) that in the particular case of one-sided (left to right) moving waves they are describable by a dissipation-modified KdV equation:

$$h_t + (h^2)_y + h_{yyy} + \delta[h_{yy} + h_{yyy} + D(h^2)_{yy} + \alpha h] = 0, \tag{5.1}$$

where $h(y,t)$ is a scaled elevation of the surface in the study of one-side steadily propagating waves. The coefficient D can be either positive or negative, while α and δ are non-negative. Equation (5.1) underlies the energy balance sustaining, say, a solitary wave. Indeed, multiplying equation (5.1) by h, and integrating over the appropriate space, if the mean value of h along the layer is zero (mass conservation), then the energy $E = \frac{1}{2} \int h^2$ is governed by the balance

$$\frac{dE}{dt} = \delta\left(\int h_x^2 \, dx - \int h_{xx}^2 \, dx + 2D \int h h_x^2 \, dx - \alpha \int h^2 \, dx\right), \tag{5.2}$$

whose value vanishes for a steadily travelling wave. The first term on the right-hand side of (5.2) describes the energy input at rather long wavelengths due to instability, the second and fourth terms describe energy dissipation on short and long wavelengths, respectively, and the third term accounts for nonlinear (feedback) correction to long-wave energy input (for h positive, positive if D is positive and negative otherwise). In the absence of dissipation and continuous energy supply, equation (5.1) reduces to the standard KdV equation and solitary waves or cnoidal waves (periodic wave trains) are still possible thanks to the disperion-nonlinearity balance also existing in equation (5.1). When dissipation plays a negligible role in experiment they can be excited from appropriate initial conditions either numerically or in water tanks where viscosity can be neglected. In the present case the situation is different. Indeed, at variance with the (integrable, dissipation-less) KdV equation where a one-parameter family of solutions exists, and hence as a consequence of initial conditions all possible amplitudes and corresponding velocities exist, the input–output energy balance (5.2) selects a single wave or a single amplitude periodic wave train, a bound state or an erratic/chaotic wave train (Christov & Velarde 1995; Nekorkin & Velade 1994; Velarde *et al.* 1995; Rednikov *et al.* 1995*b*).

On the other hand, when considering the three-dimensional problem (Nepomnyashchy & Velarde 1994) phase shifts following collision or reflection at walls depend upon the incident angle, α_i (e.g. measured front-to-front or twice the value front-to-wall, i.e. by $\frac{1}{2}\pi - \alpha_i$; a reflection is like a collision with a mirror image wave). At the approximate value of $\frac{1}{2}\pi$ no phase shift is expected while for lower collision angles the phase shift has the sign of phase shifts upon head-on collisions. Higher values than $\frac{1}{2}\pi$ (or $\alpha_i < \frac{1}{4}\pi$) lead to a change of sign in the phase shift and the formation of a Mach stem, i.e. a third wave evolving phase locked with the post collision or reflected front, a phenomenon discovered a century ago by Russell (1885) for water waves and by Mach (Courant & Friedrichs 1948; Krehl & van der Geest 1991) for shocks in gases. The phase shift sign in such case is the same as the sign in the overtaking collisions discussed by Zabusky & Kruskal (1965) in their seminal paper where they also introduced the soliton concept. Phase shifts and the formation of bound states have been numerically observed. Starting with, for example, an initial condition of two nearby 'solitary' pulses, the system evolves according to equation (5.1) to a wave train with unequally spaced maxima. All maxima have the same value, and hence the same velocity dictated by the energy balance (5.2) in the steady state.

(ii) *Experiments*

Both mass absorption and desorption, and heat transfer experiments have been carried out with Bénard layers (Weidman *et al.* 1992; Linde *et al.* 1993*a*, *b*, 1997; Wierschem 1997). As a matter of fact, with the theory sketched in the previous subsection I unearthed some 25-year-old experiments by H. Linde, who at about the same time of the numerical discovery of the soliton (Zabusky & Krustal 1965) made its experimental finding (including the collision kinematics of solitons) in surface tension gradient-driven flows.

For the case of heat transfer (liquid depths from 0.3 to 0.8 cm) liquid octane was poured in a square or cylindrical vessel, and in an annular channel (1.5 and 2.0 cm inner and outer radii, respectively). The bottom was cooled by air or water at 20 °C and the quartz cover, placed at 0.3 cm above the liquid, was heated, establishing a temperature gradient in the liquid layer. For values of this gradient ranging from 10 to 200 K cm^{-1}, solitary waves and periodic wave trains have been observed showing behaviour in head-on and oblique collisions similar to results from mass transfer experiments.

For mass transfer the following set-up was devised. In a vessel A, either a cylindrical container or an annular channel was filled with liquid (liquid depth 1.8 cm). Two vessels B$_1$ and B$_2$ were also filled with another liquid. With pentane in B$_1$ and B$_2$ either xylene, nonane, trichloroethylene or benzene were used as an absorbing liquid in A, while with toluene as an absorbing liquid in A either hexane, pentane, acetone or diethylether were used in B$_1$ and B$_2$. In all cases the results were qualitatively the same. A glass cover, C, was placed on top of the vessels B$_1$ and then when the vessel C was full of hexane vapor, say, it was placed on top of A thus allowing the absorption of hexane by the toluene liquid in A. The adsorption and subsequent absorption processes were rather strong, creating Marangoni stresses high enough to trigger and sustain instability. During the whole duration of the experiment, hexane vapor was also allowed to diffuse from the two vessels B$_2$ to A.

Observation and recording with a CCD camera was made by shadowgraph from the top with pointlike illumination from the bottom up (work is in progress with more sophisticated means). For instance, with cylindrical or annular cylindrical containers, rather violent chaotic motions occurred at first along the surface in A with waves moving in practically all directions, but finally after about one minute, when most of the vapor in C has been absorbed, a dramatic self-organization led to regular wave motion. Long-time-lasting synchronically colliding counter-rotating periodic wave trains have been observed for about 50–200 s, while the single (periodic) wave train with either clockwise or counter-clockwise rotation remained up to 450 s. As the Marangoni stresses decay the number of crests in each train increases with corresponding wavelength decrease. Subsequently, a single set of either clockwise or counter-clockwise (periodic) rotating waves remains until finally when the constraint fades away all convective motions disappear. Typical mean wave velocities at the outer wall of the annular channel before and after collision are, respectively, 2.7 and 1.7 cm s^{-1} (corresponding to angular speeds of 71.4 and 45.7° s^{-1}, respectively). Thus the mean wave velocity right after collision is about 64% (with less than 2% error) the mean wave speed measured before collision. About 0.2 s after collision the original wave speed is recovered. The phase shift shows the same sign as in the case of head-on collisions of 'solitary' waves in rectangular vessels. Reflections at walls also illustrate the solitonic/shock behaviour of the waves which occur with and without the formation of a (phase locked, third wave) Mach stem according to the angle of incidence. I have sketched just some of our findings. The phenomena observed are complex and only recently have there been clear-cut distinction between mostly surface waves (Marangoni–Bénard problem) and (mostly) internal waves (Rayleigh–Marangoni problems), all of them are triggered and sustained by the Marangoni effect, as heuristically discussed in § 5 a. Further details about the experiments can be found in Weidman *et al.* (1992), Linde *et al.* (1993*a, b*, 1997) and Wierschem (1997).

This paper summarizes work done in the past few years in my laboratory at the Instituto Pluridiciplinar in Madrid. It is the result of fruitful collaborative research with Michèle Adler, J. Bragard, C. I. Christov, A. de Ryck, H. Linde, A. A. Nepomnyashchy, A. Ye. Rednikov, M. Santiago-Rosanne, V. Starov and A. Wierschem. The research was supported by various grants: DGICYT PB93-0081, EU Network ERBCHRX-CT96-0107, EU Network ERBFMRX-CT96-0010, and Fundación BBV (Programa Cátedra, University of Cambridge).

References

Barton, K. D. & Subramanian, R. S. 1989 *J. Coll. Interf. Sci.* **133**, 211.

Borgas, M. S. & Grotberg, J. B. 1988 *J. Fluid Mech.* **193**, 151.

Bragard, J. & Velarde, M. G. 1997 *J. Non-Equilib. Thermodyn.* **19**, 95.

Bragard, J. & Velarde, M. G. 1998 *J. Fluid Mech.* (In the press.)

Bragard, J., Pontes, J. & Velarde, M. G. 1996 *Int. J. Bifur. Chaos* **6**, 1665.

Bratukhin, Yu. K., Evdokimova, O. A. & Pschenichnikov, A. F. 1979 *Izv. Akad. Nauk USSR* (*Mech. Liquids Gases*) **5**, 55–57. (In Russian.)

Bratukhin, Yu. K., Briskman, V. A., Zuev, A. L., Pschenichnikov, A. F. & Rivkind, V. Ya. 1982 In *Hydrodynamics and heat-mass transfer in weightlessness*, pp. 98–109. Moscow: Nauka. (In Russian.)

Braun, B., Ikier, Ch., Klein, H. & Woermann, D. 1994 In *Rocket experiments in fluid science and materials sciences*, vol. 4, p. 119. Paris: ESA.

Christov, C. I. & Velarde, M. G. 1995 *Physica D* **86**, 323.

Chu, X.-L. & Velarde, M. G. 1988 *Physico Chem. Hydrodyn.* **10**, 727.

Chu, X.-L. & Velarde, M. G. 1991 *Phys. Rev.* A **43**, 1094.

Courant, R. & Friedrichs, K. O. 1948 *Supersonic flow and shock waves*. New York: Interscience.

Davis, S. H. 1987 *A. Rev. Fluid Mech.* **19**, 403.

Edwards, D. A., Brenner, H. & Wasan, D. T. 1991 *Interfacial transport processes and rheology* Boston, MA: Butterworth-Heineman.

Garazo, A. N. & Velarde, M. G. 1991 *Phys. Fluids* A **3**, 2295.

García-Ybarra, P. L. & Velarde, M. G. 1987 *Phys. Fluids* **30**, 1649.

Gaver, D. P. & Grotberg, J. B. 1990 *J. Fluid Mech.* **213**, 127.

Gaver, D. P. & Grotberg, J. B. 1992 *J. Fluid Mech.* **235**, 399.

Happel, J. & Brenner, H. 1965 *Low Reynolds number hydrodynamics*. Englewood Cliffs, NJ: Prentice-Hall.

Hardy, S. C. 1979 *J. Coll. Interf. Sci.* **69**, 157.

Koschmieder, E. L. 1993 *Bénard cells and Taylor vortices*. Cambridge University Press.

Krehl, P. & van der Geest, M. 1991 *Shock Waves* **1**, 3.

Levchenko, E. B. & Chernyakov, A. L. 1981 *Sov. Phys. JETP* **54**, 102.

Levich, B. G. 1965 *Physicohemical hydrodynamics*. Englewood Cliffs, NJ: Prentice-Hall.

Levich, B. G. & Krylov, V. S. 1969 *A. Rev. Fluid Mech.* **1**, 293.

Linde, H., Chu, X.-L. & Velarde, M. G. 1993*a* *Phys. Fluids* A **5**, 1068.

Linde, H., Chu, X.-L., Velarde, M. G. & Waldhelm, W. 1993*b* *Phys. Fluids* A **5**, 3162.

Linde, H., Velarde, M. G., Wierschem, A., Waldhelm, W., Loeschcke, K. & Rednikov, A. Ye. 1997 *J. Coll. Interf. Sci.* **188**, 16.

Lucassen, J. 1968 *Trans. Faraday Soc.* **64**, 2221.

Manneville, P. 1990 *Dissipative structures and weak turbulence.* San Diego, CA: Academic.

Merritt, R. M. & Subramanian, R. S. 1988 *J. Coll. Interf. Sci.* **125**, 333.

Nekorkin, V. I. & Velarde, M. G. 1994 *Int. J. Bifur. Chaos* **4**, 1135.

Nepomnyashchy, A. A. & Velarde, M. G. 1994 *Phys. Fluids* **6**, 187.

Normand, Ch., Pomeau, Y. & Velarde, M. G. 1977 *Rev. Mod. Phys.* **49**, 581, and references therein.

Ostrach, S. 1982 *A. Rev. Fluid Mech.* **14**, 313.

Pearson, J. R. J. 1958 *Fluid Mech.* **4**, 489.

Proudman, I. & Pearson, J. R. A. 1957 *J. Fluid Mech.* **2**, 237.

Rednikov, A. Ye., Ryazantsev, Yu. S. & Velarde, M. G. 1994*a Phys. Fluids* **6**, 451.

Rednikov, A. Ye., Ryazantsev, Yu. S. & Velarde, M. G. 1994*b J. Coll. Interf. Sci.* **164**, 168.

Rednikov, A. Ye., Ryazantsev, Yu. S. & Velarde, M. G. 1994*c Int. J. Heat. Mass Transf.* **37**, supp. 1, 361.

Rednikov, A. Ye., Ryazantsev, Yu. S. & Velarde, M. G. 1994*d J. Non-Equilib. Thermodyn.* **19**, 95.

Rednikov, A. Ye., Ryazantsev, Yu. S. & Velarde, M. G. 1994*e Phys. Scripta* T **55**, 115.

Rednikov, A. Ye., Kurdyumov, V. N., Ryazantsev, Yu. S. & Velarde, M. G. 1995*a Phys. Fluids* **7**, 2670.

Rednikov, A. Ye., Velarde, M. G., Ryazantsev, Yu. S., Nepomnyashchy A. A. & Kurdyumov, V. N. 1995*b Acta Appl. Math.* **39**, 457.

Rednikov, A. Ye., Colinet, P., Velarde, M. G. & Legros, J. C. 1998*a* Oscillatory themocapillary instability in a liquid layer with deformable open surface: capillary-gravity waves, longitudinal waves and mode-mixing. *J. Fluid Mech.* (Submitted).

Rednikov, A. Ye., Colinet, P., Velarde, M. G. & Legros, J. C. 1998*b* Rayleigh–Marangoni oscillatory instability in a liquid layer heated from above: coupling between internal and surface waves. *J. Fluid Mech.* (Submitted).

Russell, J. S. 1885 *The wave of translation in the oceans of water, air and ether.* London: Trübner & Co.

Santiago-Rosanne, M., Adler, M. & Velarde, M. G. 1997 *J. Coll. Interf. Sci.* **191**, 65.

Scriven, L. E. & Sternling, C. V. 1960 *Nature* **187**, 186.

Starov, V., Ryck, A. de & Velarde, M. G. 1997 *J. Coll. Interf. Sci.* **190**, 104.

Subramanian, R. S. 1981 *AIChE. Jl* **27**, 646.

Subramanian, R. S. 1992 In *Transport processes in bubbles, drops and particles* (ed. R. P. Chhabra & D. de Kee). New York: Hemisphere.

Taylor, T. D. & Acrivos, A. 1964 *J. Fluid. Mech.* **18**, 466.

Velarde, M. G. & Normand, C. 1980 *Sci. Am.* **243**, 92.

Velarde, M. G., Chu, X.-L. & Garazo, A. N. 1991 *Phys. Scripta* T **35**, 71.

Velarde, M. G., Nekorkin, V. I. & Maksimov, A. 1995 *Int. J. Bifur. Chaos* **5**, 831.

Velarde, M. G., Rednikov, A. Ye. & Ryazantsev, Yu. S. 1996 *J. Phys.* C **8**, 9233.

Weidman, P. D., Linde, H. & Velarde, M. G. 1992 *Phys. Fluids* A **4**, 921.

Wierschem, A. 1997 Ph.D. dissertation, Humboldt Universitat, Berlin.

Wozniak, G., Siekmann, J. & Srulijes, J. 1988 *Z. Flugwiss. Weltraumforsch.* **12**, 137.

Young, N. O., Goldstein J. S. & Block, M. J. 1959 *J. Fluid Mech.* **6**, 350.

Zabusky, N. J. & Kruskal, M. D. 1965 *Phys. Rev. Lett.* **15**, 57.

Discussion

S. K. WILSON (*Department of Mathematics, University of Strathclyde, Glasgow, UK*). As an applied mathematician working on the theoretical investigation of

Marangoni phenomena it has been very interesting to see some new situations in materials processing in which Marangoni effects are important. However, researchers in materials processing should be aware that work on similar phenomena is ongoing in other areas. These include crystal growing (about which we have heard something at this Meeting), the coating industry (see, for example, the paper by Wilson (1993)) and biological applications (see, for example, the recent review article by Grotberg (1994)). In all these cases I'm sure both parties would benefit from collaboration between workers in apparently very different subject areas.

J. R. HELLIWELL (*Department of Chemistry, University of Manchester, Manchester, UK*). Marangoni convection driven fluid flow is one type of convection effect. In any 'real' experiment, such as one to produce a high quality protein crystal (a type of materials processing, certainly), there are a variety of effects which can defeat the reaching of the ideal situation, which I define here as growing a protein crystal that does not move in the fluid due to sedimentation (on Earth), or due to g-jitter in orbit, and where also the mother liquor is not subject to (turbulent) fluid flow. A real situation then is a superposition of perturbation effects away from an ideal and each of which may induce some sort of defect. Neverthless, the choice of particular crystallization geometry, e.g. avoiding vapour diffusion (i.e. liquid–vapour interface) can avoid Marangoni effects. However, in a linear liquid–liquid diffusion geometry other effects due to g-jitter are more difficult to eliminate and, except perhaps in the case of an uncrewed orbiting platform, may always prevent the ideal conditions every truly being realized. I am interested to hear from other areas of 'materials processing' the relative importance for Marangoni over other convection or g-jitter driven effects. Perhaps cases like crystal growth form a liquid bridge, of initially molten material, simply cannot avoid Marangoni convection? By contrast, in protein crystal growth it seems that Marangoni convection can be avoided rather simply by avoidance of liquid vapour droplet geometry. The exploration of the ultimate protein crystal quality obtainable relates to the exceptionally fine X-ray brilliance available from the new generation of SR X-ray sources and harnessing their full technical capability. In addition, the overall motivation for such studies in protein crystallography is to better understand the factors that can lead to a poor quality crystal or indeed no crystal at all (or at best a 'microcrystal').

Additional references

Grotberg, J. B. 1994 Pulmonary flow and transport phenomena. *A. Rev. Fluid Mech.* **26**, 529–571.

Wilson, S. K. 1993 The levelling of paint films. *IMA J. Appl. Math.* **50**, 149–166.

Measurements of thermophysical properties of liquid metals relevant to Marangoni effects

By I. Egry, M. Langen and G. Lohöfer

Institut für Raumsimulation, DLR, 51170 Köln, Germany

Marangoni convection is caused by a gradient in the surface tension along a free liquid surface. The dimensionless Marangoni number, which controls the strength of this convection, contains additional thermophysical parameters. For liquid metals, these quantities are best measured under containerless conditions using electromagnetic levitation and non-contact diagnostic tools. In microgravity, small electromagnetic fields are sufficient to position a liquid sample. Some experiments can only be performed in such an environment, most others greatly benefit from microgravity and lead to results of higher precision. This paper reports on both terrestrial and microgravity measurements of thermophysical properties of undercooled liquid metals, including specific heat, density, surface tension, viscosity and electrical conductivity.

Keywords: oscillating drop technique; electromagnetic levitation; surface tension; density; electrical conductivity; microgravity

1. Introduction

Marangoni convection is caused by a gradient in the surface tension γ along a free liquid–vapour interface. This driving force must overcome the resistance of the fluid to flow, characterized by the viscosity η. The dimensionless Marangoni number, Ma, expresses this competition. It is defined as

$$Ma = (L_s \rho c_p \delta\gamma)/(\lambda\eta), \tag{1.1}$$

where L_s is a characteristic length, ρ is the density, c_p is the specific heat, $\delta\gamma$ is the difference in surface tension along L_s, and λ is the thermal conductivity. In order to accurately predict the flow pattern, the thermophysical parameters entering into the definition of the Marangoni number must be precisely known. This is a formidable task, because it involves the measurement of five different parameters. A particular difficulty arises from the determination of $\delta\gamma$. In most cases, this difference is due to a temperature gradient δT along L_s, and it can therefore be written as

$$\delta\gamma = \partial\gamma/\partial T\, \delta T. \tag{1.2}$$

Therefore, the temperature coefficient of the surface tension has to be determined, which involves the differentiation of the primarily determined surface tension $\gamma(T)$ with respect to temperature. Accurate results can only be obtained if a wide temperature range is covered and the scatter of the original data points is small.

Inserting equation (1.2) into (1.1), we can separate external quantities from intrinsic material properties by writing

$$Ma = L_s\delta T\, ma, \quad ma = (\rho c_p \partial\gamma/\partial T)/(\lambda\eta). \tag{1.3}$$

31

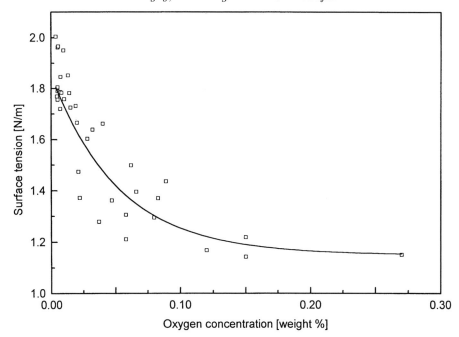

Figure 1. Surface tension of iron as a function of oxygen concentration at 1650 °C (after Keene
et al. 1982).

We are concerned with the determination of the specific Marangoni number, *ma*. Once this quantity is known, *Ma* for a given experimental setup can be obtained easily from equation (1.3). Note that *ma* can be positive or negative, depending on the sign of $\partial\gamma/\partial T$. The temperature coefficient $\partial\gamma/\partial T$ should be negative for pure elements (with the possible exception of Ga) due to the decrease of structural differences between liquid and gas with increasing temperature; note that, at the critical temperature T_c, both phases become equal, and, consequently, their interface disappears. However, for alloys, segregation effects may become dominant at low temperatures, leading to a decrease of $\partial\gamma/\partial T$ with decreasing temperature and thereby to a positive temperature coefficient.

Whereas reliable data exist for fluids which are liquid at, or slightly above, room temperature, the situation is different for high-temperature melts, like liquid metals with a melting temperature around 1000 °C, typically. At these high temperatures 'everything reacts with everything' (Mills & Brooks 1994), and it is difficult to find a container that does not contaminate the specimen under investigation. To complicate matters, surface tension is particularly sensitive to even small amounts of impurities. As an example, the surface tension of iron is shown as a function of oxgen concentration in figure 1 (Keene *et al.* 1982).

For liquid metals, electromagnetic levitation provides containerless processing capabilities. An inhomogeneous RF electromagnetic field exerts a Lorentz force on a metallic sample and lifts it against gravity. The ohmic losses of the induced eddy currents in the sample heat, and eventually melt, the sample. If non-contact diagnostic tools can be developed which are compatible with electromagnetic levitation, this combination is best suited for the study of liquid metals. Recently, considerable progress has been made in this direction (Egry *et al.* 1993). Containerless processing has the additional advantage that the liquid metals can be easily undercooled:

due to the absence of container walls, the number of heterogeneous nucleation sites is greatly reduced, and nucleation is delayed. There is, however, one shortcoming of electromagnetic levitation: the electromagnetic fields not only lift and heat the sample, but they also deform its shape and induce potentially turbulent flows in the sample. These undesired side effects cannot be eliminated on Earth and can only partially be accounted for by some extensive magnetohydrodynamic calculations (Cummings & Blackburn 1991; Suryanarayana & Bayazitoglou 1991; Bratz & Egry 1995). For this reason, experiments under microgravity conditions are useful and have been performed (Team TEMPUS 1996).

This paper reviews the non-contact experimental techniques available today in combination with electromagnetic levitation. They allow us to measure all of the thermophysical properties entering the Marangoni number, namely density, specific heat, surface tension, viscosity, and, indirectly, thermal conductivity. A discussion of the results obtained so far, including microgravity experiments, is also included.

2. Specific heat

The specific heat c_p describes the temperature change of a body due to heat input or heat loss:

$$mc_p \frac{dT}{dt} = P_{\text{in}} - P_{\text{out}}, \tag{2.1}$$

where m is the mass of the body and P_{in} and P_{out} are the power input and output, respectively. In electromagnetic levitation, power is fed into the sample by induction. The power absorbed by the sample is proportional to the power drawn by the levitation coil. Once this coupling coefficient is known, the power absorbed by the sample can be calculated from the power draw of the RF circuit. To determine the specific heat, the heat loss must also be known. This is most easily achieved in vacuum conditions, where there are only radiative heat losses:

$$P_{\text{out}} = \sigma_{\text{SB}} A \epsilon T^4, \tag{2.2}$$

where σ_{SB} is the Stefan–Boltzmann constant, A is the surface area of the specimen, and ϵ is the total hemispherical emissivity. In terrestrial levitation, convective cooling is always required to control and limit the sample temperature. The effect of the cooling gas has to be taken into account in the heat balance (equation (2.1)), and, if convection is present, the heat loss to the gas cannot be simply described by an effective thermal conductivity. Therefore, the applicability of this method is limited to microgravity experiments.

If the total hemispherical emissivity is known, the specific heat can be obtained from cooling curves, i.e. $P_{\text{in}} = 0$. In such a case, c_p is simply given by

$$c_p = -\frac{\sigma_{\text{SB}} A \epsilon}{m} \frac{T^4}{dT/dt}. \tag{2.3}$$

Unfortunately, in most cases, ϵ is not known. Nevertheless, the specific heat can be determined through a modulation technique, developed by Fecht & Johnson (1991). The heater power is modulated according to $P(t) = P_\omega \cos(\omega t)$ resulting in a modulated temperature response ΔT_ω of the sample. Temperature gradients inside the sample relax quickly, due to the high thermal conductivity of metals. This relaxation can be described by a relaxation time τ_{int}. On the other hand, relaxation to the equilibrium temperature is governed by radiation under UHV conditions, and is therefore

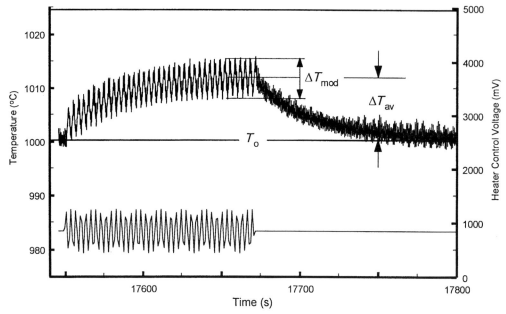

Figure 2. AC calorimetry on ZrNi alloy: power modulation and temperature response

slow. It can be described by a relaxation time τ_{ext}. If the modulation frequency ω is chosen such that $1/\tau_{\text{ext}} \ll \omega \ll 1/\tau_{\text{int}}$, a simple relation for the temperature variation can be derived:

$$\Delta T_\omega = P_\omega/(\omega c_p), \tag{2.4}$$

from which c_p can be determined.

In practice, the power input into the sample is controlled by a control voltage U_{c}:

$$P_{\text{in}} = \alpha U_{\text{c}}^2, \tag{2.5}$$

where α is a constant characterizing the RF circuit including the sample. The control voltage is modulated according to

$$U_{\text{c}} = U_0 + U_{\text{m}} \cos(\omega t), \tag{2.6}$$

resulting in a power modulation at the sample:

$$P_{\text{in}} = \alpha(U_0^2 + \tfrac{1}{2}U_{\text{m}}^2 + 2U_0 U_{\text{m}} \cos(\omega t) + \tfrac{1}{2}U_{\text{m}}^2 \cos(2\omega t)). \tag{2.7}$$

The first term is the unmodulated RF power. It determines the equilibrium temperature of the sample before modulation. As we can see from (2.7), the modulation produces a periodic temperature response (third and fourth terms) superimposed over an increase in the average sample temperature (second term). Equation (2.4) can be applied to the third and fourth term separately. Figure 2 shows data taken from an $Ni_{24}Zr_{76}$ sample flown on IML-2 (Team TEMPUS 1996) with the corresponding heater control voltage plotted below.

Of course, specific heat data from liquid, and even undercooled metals, can also be measured using drop calorimeters (Barth *et al.* 1993). The fundamental problem of drop calorimetry is due to the fact that the specific heat is not measured directly, but is obtained by differentiating the measured enthalpy with respect to temperature. To discuss these measurements in any detail is outside the scope of the present article.

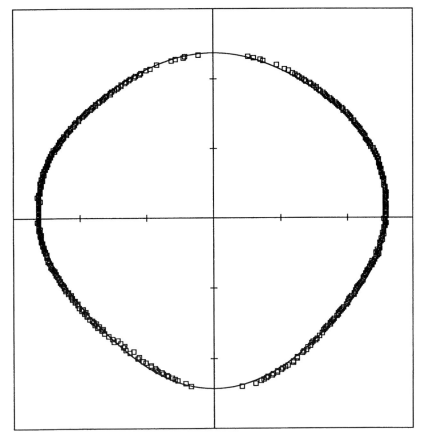

Figure 3. Shape of a levitated silicon sample and fit with Legendre polynomials.

3. Density

Density measurements of levitated samples can be made using videography. In terrestrial levitation, samples are not spherical, but slightly elongated due to the action of gravity and the electromagnetic field. However, their static equilibrium shape is still rotationally symmetrical around the vertical axis (parallel to the gravity vector). Therefore, images are taken perpendicular to this axis, and the volume V of a rotationally symmetrical body is calculated. The mass m of the sample is known; it is weighed before and after the measurement. The density of the sample is then obtained from

$$\rho = m/V. \tag{3.1}$$

The images are taken at constant temperature and analysed off-line by a digital image processing system. In a first step, the software detects the edge of the incandescent sample; then an average of approximately 100 images is performed to remove the potentially asymmetrical dynamic surface oscillations. Finally, the shape of the averaged image is fitted with a series of Legendre polynomials.

An example of such a fit is shown in figure 3.

Once the coefficients of this series expansion are known, the volume and hence the density can be calculated. A detailed description of this algorithm has been given in

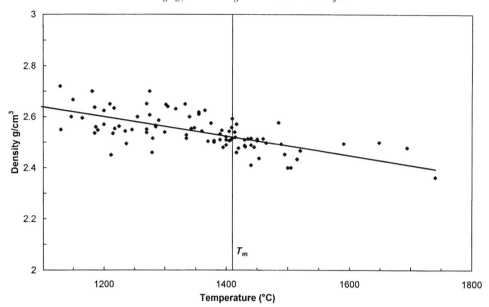

Figure 4. Density of liquid silicon as function of temperature.

Gorges *et al.* (1996). Using this method, we have recently determined the density of liquid silicon over a wide temperature range. The result is shown in figure 4.

This example shows that it is even possible to apply electromagnetic levitation to semiconductors.

4. Surface tension and viscosity

(*a*) *Surface tension*

The oscillating drop technique is an elegant way to measure both surface tension and viscosity. It employs digital image processing for frequency analysis of surface waves. The radius a of a droplet undergoes oscillations of the form

$$\delta a_{l,m}(\vartheta, \phi, t) \propto Y_{l,m}(\vartheta, \phi) \cos(\omega_{l,m} t) e^{-\Gamma_{l,m} t}. \tag{4.1}$$

Here, $Y_{l,m}$ are spherical harmonics. The frequency $\omega_{l,m}$ is related to surface tension, while the damping $\Gamma_{l,m}$ of the waves is due to viscosity. If the equilibrium shape of the droplet is spherical, the simple formulae of Rayleigh and Kelvin can be used to relate frequency ω and damping Γ of the oscillations to surface tension γ and viscosity η, respectively. Rayleigh's formula reads

$$\omega_{\mathrm{R}}^2 = \tfrac{32}{3}\pi(\gamma/m), \tag{4.2}$$

while Kelvin derived the following expression:

$$\Gamma_{\mathrm{K}} = \tfrac{20}{3}\pi(a\eta/m), \tag{4.3}$$

where m is the mass of the droplet and a is its radius. These two expressions relate to the fundamental mode of oscillation, which corresponds to $l = 2$. For spherical drops, frequencies and damping constants do not depend on m ($|m| < 2$). A spherical shape is obtained only if the droplet is free of external forces. This situation is well approximated in microgravity. Under terrestrial conditions, the above relations are

Table 1. *Surface tension of liquid transition and noble metals*

element	γ_m (mN m^{-1})		γ_T (mN m^{-1} K^{-1})	
	this work	Keene	this work	Keene
Fe	1870	1862	0.43	0.39
Co	1874	1881	0.3	0.34
Ni	1770	1796	0.33	0.35
Cu	1304	1330	0.22	0.23
Ag	908	925	0.18	0.21
Au	1149	1145	0.14	0.18

not valid and corrections have to be made for the external forces, namely gravity and electromagnetic field. These corrections have been calculated recently (Cummings & Blackburn 1991; Suryanarayana & Bayazitoglou 1991; Bratz & Egry 1995). They take into account both the splitting of the peaks due to symmetry breaking, and the shifting of the peaks due to magnetic pressure. For the Rayleigh formula the correction reads

$$\tfrac{32}{3}\pi(\gamma/m) = \tfrac{1}{5}\sum_m \omega_{2,m}^2 - 1.9\overline{\Omega_{\mathrm{tr}}^2} - 0.3(\overline{\Omega_{\mathrm{tr}}^2})^{-4}(g/a)^2 \qquad (4.4)$$

Here, $\overline{\Omega_{\mathrm{tr}}^2}$ is the mean of the translational frequencies of the sample in the potential well of the levitation field, and g is the gravitational acceleration. It has been shown that by applying the Cummings correction to surface tension data obtained by the oscillating drop technique on Earth, a spurious mass dependence can be eliminated (Egry 1994). For gold, the value thus obtained agrees well with data obtained using the sessile drop technique. In addition, Egry and coworkers have performed microgravity experiments on gold and a gold–copper alloy (Egry *et al.* 1995). These experiments clearly show a single peak in the oscillation spectrum which means that the frequencies do not depend on m and, furthermore, they yield values for the surface tension which are in excellent agreement with terrestrial data only if the latter are corrected according to equation (4.4).

Using the oscillating drop method and applying the Cummings correction, we have measured the surface tension of a number of liquid metals. The temperature dependence of the surface tension of pure elements is conveniently described by

$$\gamma(T) = \gamma_m - \gamma_T(T - T_{\mathrm{m}}), \qquad (4.5)$$

where T_{m} is the melting temperature. Our results for noble and transition metals are listed in table 1. For comparison, the recommended values of Keene (1993) are also shown. Generally speaking, the agreement is excellent, with our data being somewhat lower. This may be due to the fact that Keene's compilation also contains data obtained with the oscillating drop technique, but without Cummings correction.

So far, we have only discussed pure elements. In the case of alloys, the surface tension depends on both temperature and concentration. Whereas the temperature dependence is essentially linear, the concentration dependence is more complicated. This is due to surface segregation effects. In alloys, the system uses its additional degree of freedom to minimize its free energy. It can do so by segregating the

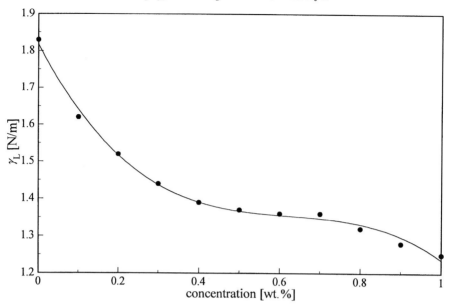

Figure 5. Surface tension of the system Cu–Ni.

component with lower surface tension at the surface. This gain in energy is partly compensated by loss of entropy, particularly at high temperatures. The surface
tension of alloys can be calculated from conventional bulk thermodynamics if the
mixing character is known. For the simplest case of an ideal solution, the following
explicit formula can be given:

$$\gamma_{\mathrm{mix}} = \gamma_1 - \tilde{\gamma}\ln\{c_1 + (1 - c_1)\exp(\Delta\gamma/\tilde{\gamma})\}, \tag{4.6}$$

where γ_1 and c_1 are the surface tension and concentration of component 1, respectively. The difference between the surface tensions of the pure components is
$\Delta\gamma = \gamma_1 - \gamma_2$, and $\tilde{\gamma} = RT/f$, where f is the surface area of one mole of either
species. Usually f is treated as a fitting parameter. More generally, the surface ten-
sion of alloys is calculated from Butler's formula assuming either regular or subregular
solutions (Hajra et al. 1991). We have measured the surface tension of the simple
system Cu–Ni (Gorges 1996). It is a completely miscible system with a very simple
phase diagram. Therefore, one should expect ideal mixing behaviour. However, as
can be seen from figure 5, CuNi does not mix ideally, and it is best described by a
regular solution model.

(b) Viscosity

In the case of viscosity, the Kelvin formula (equation (4.3)) is derived for force-
free samples under the assumption of purely laminar flow. Whereas the effect of
external forces on the equilibrium shape can be taken into account (as long as they
can be treated as a small perturbation), this is certainly not true when they cause
turbulent flow. In such a case, turbulence introduces additional damping which masks
the damping due to viscosity. This seems to be the case for both terrestrial and
microgravity electromagnetic levitation. Therefore, until now, no viscosity data could
be derived from the oscillating drop technique. If the levitation fields can be further
reduced, there is hope that this method can be applied to high-viscosity systems
such as PdCuSi or the recently discovered easy glass formers (Johnson 1996; Inoue

et al. 1994). Corresponding microgravity experiments are under way: a reflight of the TEMPUS facility is planned for SpaceHab mission MSL-1 in July 1997. Conventional methods like the oscillating vessel or oscillating bob viscometers are difficult to use for liquid metals and can have errors of up to 50% (Iida & Guthrie 1993). One way of at least estimating viscosity values in the liquid and undercooled regime is provided by the simple formula (Egry 1993):

$$\gamma/\eta = \tfrac{15}{16}\sqrt{kT/m}.$$ (4.7)

5. Conductivity

Finally, it is also possible to measure the electrical conductivity σ of levitated droplets. This can be done using non-contact inductive methods. The basic idea is to place a pickup coil around the sample and to measure the impedance Z of the system (Lohöfer 1994). Being a complex quantity, Z contains information about the ratio of the amplitudes of voltage and current, U_0/I_0, as well as their phase shift ϕ. Any changes in the impedance can be attributed to changes of the sample. Two effects can influence the impedance, namely the sample can either change its conductivity or its shape. For conductivity measurements it is essential that these two effects can be separated. It can be shown that for small skin depth δ and small deviations from a spherical shape, this is indeed the case, if the pickup coil has a special geometry. The skin depth δ is defined as

$$\delta = \sqrt{2/\omega\mu_0\sigma}.$$ (5.1)

Here, ω is the oscillation frequency of the generator circuit, μ_0 is the magnetic permeability and σ is the conductivity. Therefore, small skin depth implies high frequency or high conductivity. In terrestrial levitation experiments, the levitation coil cannot be used as a pickup coil for impedance measurements and an additional measuring coil has to be introduced. In microgravity experiments, the heating coil has the required symmetry and can be used as pickup coil. In addition, the liquid sample is spherical. For such a geometry, a simple relation can be derived between U_0/I_0 and δ, which allows the determination of the conductivity. It reads

$$\delta = \frac{a}{2}\left(1 - \sqrt{1 - 4\left\{\frac{A}{U_0/I_0} - B\right\}}\right).$$ (5.2)

A and B are two constants characterizing the empty coil and a is the radius of the sample. During the SpaceHab mission IML-2, the electrical conductivity of a $Zr_{64}Ni_{36}$ alloy was measured in the TEMPUS facility using the approach outlined above. Figure 6 shows the result.

Presently, we are developing modified coil systems and an improved theory which will allow us to measure the electrical conductivity of slightly deformed samples on Earth.

To obtain the thermal conductivity λ, needed to complete the definition of the specific Marangoni number ma, we have to use the Wiedemann–Franz law which relates thermal and electrical conductivity (Iida & Guthrie 1993):

$$\lambda/\sigma = LT,$$ (5.3)

where L is the Lorenz number: $L = 2.45 \times 10^8$ V^2 K^{-2}. The Wiedemann–Franz law is known to hold well for metals at high temperatures.

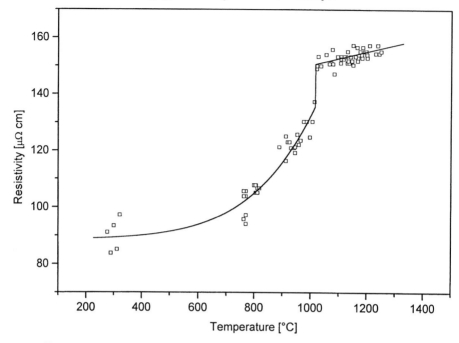

Figure 6. Electrical resistivity of $Zr_{64}Ni_{36}$ in the solid and liquid phase.

6. Conclusion

The measurement of all thermophysical properties determining the Marangoni number is a difficult task, in particular for liquid metals at high temperatures. A promising strategy is to use containerless methods which avoid contamination of the sample. Non-contact diagnostic tools are available or are being developed. With their help, the database on thermophysical properties of liquid metals will be expanded and consolidated. Some of the presented experiments rely on a microgravity environment and can therefore not yet be performed routinely. With the advent of a permanent space station, frequent experimental campaigns should become possible improving the accuracy and fidelity of the data.

References

Barth, M., Joo, F., Wei, B. & Herlach, D. M. 1993 *J. Non-crystal. Solids* **156–158**, 398.

Bratz, A. & Egry, I. 1995 *J. Fluid Mech.* **298**, 341.

Cummings, D. & Blackburn, D. 1991 *J. Fluid Mech.* **224**, 395.

Egry, I. 1993 *Scripta Metal. Mater.* **28**, 1273.

Egry, I. 1994 *Mat. Sci. Engng* A **178**, 73.

Egry, I., Lohöfer, G. & Sauerland, S. 1993 *Int. J. Thermophys.* **14**, 573.

Egry, I., Lohöfer, G. & Jacobs, G. 1995 *Phys. Rev. Lett.* **75**, 4043.

Fecht, H. & Johnson, W. 1991 *Rev. Sci. Instrum.* **62**, 1299.

Gorges, E. 1996 Ph.D. thesis, University of Aachen.

Gorges, E., Racz, L. M., Schillings, A. & Egry, I. 1996 *Int. J. Thermophys.* **17**, 1163.

Hajra, J. P., Lee, H.-K. & Frohberg, M. G. 1991 *Z. Metal.* **82**, 603.

Iida, T. & Guthrie, R. I. L. 1993 *The physical properties of liquid metals*. Oxford: Clarendon.

Inoue, A., Kawase, D., Tsai, A. P., Zhang, T. & Masumoto, T. 1994 *Mat. Sci. Engng* A **178**, 255.

Johnson, W. L. 1996 *Materials science forum*, vol. 35, pp. 225–227. Switzerland: Transtech Publications.

Keene, B. J. 1993 *Int. Mat. Rev.* **38**, 157.

Keene, B. J., Mills, K. C., Bryant, J. W. & Hondros, E. D. 1982 *Can. Met. Q* **21**, 393.

Lohöfer, G. 1994 *Int. J. Engng Sci.* **32**, 107.

Mills, K. C. & Brooks, R. F. 1994 *Mat. Sci. Engng* A **178**, 77.

Suryanarayana, P. V. R. & Bayazitoglou, Y. 1991 *Phys. Fluids* A **3**, 967.

Team TEMPUS 1996 *Materials and fluids under low gravity* (ed. L. Ratke, H. Walter & B. Feuerbacher), p. 233. Berlin: Springer.

Discussion

J. C. EARNSHAW (*Department of Pure and Applied Physics, University of Belfast, UK*). The elegant and non-invasive methods outlined in this paper promise very clean data on thermophysical properties of liquid metals. I am, however, somewhat concerned about certain effects which may influence the values of viscosity determined from the damping of the oscillations of a liquid metal drop.

Simulations have suggested that density oscillations exist at the surface of a liquid metal, reflecting the variation of the conduction electron density through the transition zone (D'Evelyn & Rice 1981). Such a layered structure at a liquid metal–vapour interface, extending some few atomic layers into the bulk, has been confirmed by X-ray reflectivity studies (Sluis & Rice 1983; Magnussen *et al.* 1995). Such a structure could act as an interfacial molecular film, supporting dilatational surface waves. Such dilatational modes would couple to the capillary modes Dr Egry observed; the main effect would be to increase the damping of the capillary modes (Lucassen-Reynders & Lucassen 1969; Kramer 1971). Indeed light scattering studies of thermally excited capillary waves on the clean surface of Hg show such increased damping (Kolevson, personal communication). The expected effects are shown in figure 7 for capillary waves of wave number $q = 10$ cm^{-1} (comparable to those in experiments by Dr Egry) on a planar Hg–vacuum interface. The considerable increase in the wave damping for non-zero dilatational elastic moduli is apparent. While the theoretical formulation is more complex for oscillations of spherical droplets (Sparling & Sedlak 1989) the two surface modes couple similarly in this case also, again leading to increased capillary mode damping. This suggests that viscosities deduced from the onserved damping values might be significantly and systematically overestimated. The changes in the capillary mode frequency are orders of magnitude less than those in the damping (see figure 7), so that estimates of surface tension should be relatively unaffected.

I. EGRY. Professor Earnshaw makes a very important remark about the interpretation of the damping of capillary waves. He points out that the damping may be due to other mechanisms than viscosity, in particular coupling to dilatational surface waves may lead to damping of the capillary waves. Strong damping of thermally excited capillary waves observed on a clean planar surface of liquid Hg was interpreted as being due to such an effect, assuming that such a surface layer is an intrinsic property of a liquid metal and would exist even on clean surfaces.

In the experiments discussed in our paper, the damping of mechanically excited oscillations of a liquid drop is observed. If surface dilatational waves existed on the surface, it is very likely that a similar mode-coupling as in the planar case would lead to an enhanced damping. However, this is not the case experimentally. Recently, we have evaluated our microgravity experiment on the eutectic $Pd_{78}Cu_6Si_{16}$ alloy, using

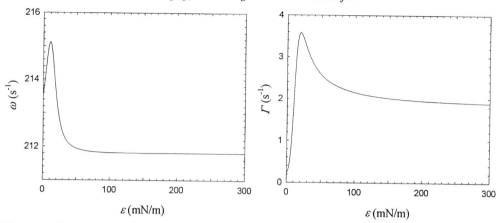

Figure 7. The capillary wave frequency (ω) and damping (Γ) for capillary waves of $q = 10 \text{ cm}^{-1}$ on a Hg–vacuum interface as a function of the dilatational elastic modulus ε. The damping varies by a factor of over 10, whereas the variations of the frequency are less than 0.7% of the value for $\varepsilon = 0$.

the simple formula $\Gamma = \frac{20}{3}\pi(a\eta/m)$ (equation (4.3) of our paper), which assumes damping by viscosity only. For the viscosity at the eutectic temperature $T = 760\,^\circ\text{C}$, our preliminary analysis yields $\eta = 49\,\text{mPa s}$ (Egry *et al.* 1998). Previously, the viscosity of $Pd_{78}Cu_6Si_{16}$ was measured by Lee *et al.* (1991). They obtained $\eta = 61\,\text{mPa s}$ at the same temperature. Our value is of the same order of magnitude, but lower than theirs. An additional damping due to the coupling to surface dilatational waves, seems therefore negligible. More work, both theoretical and experimental, is needed to clarify this puzzle.

Additional references

Egry, I., Lohöfer, G., Seyhan, I. & Schneider, S. 1998 (In the press.)

D'Evelyn, M. P. & Rice, S. A. 1981 *Faraday Symp. Chem. Soc.* **16**, 71.

Kramer, L. 1971 *J. Chem. Phys* **55**, 2097.

Lee, S. K., Tsang, K. H. & Kui, H. W. 1991 *J. Appl. Phys.* **79**, 4842.

Lucassen-Reynders, E. H. & Lucassen, J. 1969 *Adv. Colloid. Interf. Sci.* **2**, 347.

Magnussen, O. M., Ocko, B. M., Regan, M. J., Penanen, K., Pershan, P. S. & Deutsch, M. 1995 *Phys. Rev. Lett.* **74**, 4444.

Sluis, D. & Rice, S. A. 1983 *J. Chem. Phys.* **79**, 5658.

Sparling, L. C. & Sedlak, J. E. 1989 *Phys. Rev. A* **39**, 1351.

Oxygen transport and dynamic surface tension of liquid metals

By Enrica Ricci[1], Lorenzo Nanni[2] and Alberto Passerone[1]

[1]Istituto di Chimica Fisica Applicata dei Materiali, MITER-CNR,
Via De Marini 6, 16149 Genova, Italy
[2]ISTIC, University of Genova, Via all'Opera Pia 15, 16145 Genova, Italy

Theoretical models have been developed to study the behaviour of liquid metal surfaces in the presence of gaseous surface active elements, relating the mass exchange between the liquid metal and the surrounding atmosphere under the Knudsen regime. Steady-state conditions can be defined for which an 'effective oxidation pressure' is formulated, providing a useful means for a sound choice of the most convenient conditions for oxygen-free high-temperature surface tension measurements. The experimental results presented here concern dynamic surface tension measurements in a time window ranging from half a second to hours. They confirm the theoretical predictions of the model and reveal unexpected behaviours of the liquid surface when subjected to different temperature profiles.

Keywords: surface tension; liquid tin; liquid metals; oxygen adsorption

1. Introduction

When an interface between two immiscible fluids is subjected to a temperature (or concentration) gradient, its interfacial tension varies from point to point: these gradients along the surface induce shear stresses that result in fluid motion. For a pure liquid in a temperature field parallel to the surface, there is a flow from the hot end towards the cold end. Since bulk fluids are viscous, they are dragged along: thus, bulk fluid motion results from interfacial temperature or concentration gradients (thermocapillary or solutocapillary effects). The velocity profile, which is a function of the vertical distance of the free surface (or the interface) from the base plane can attain a maximum velocity ranging from some $mm\,s^{-1}$ to several $cm\,s^{-1}$, depending on the surface tension temperature coefficient and on the liquid viscosity.

The phenomena arising from interfacial tension gradients are collectively termed 'Marangoni effects', after the Italian physicist Carlo Marangoni (1840–1925) who published a series of papers on this subject between 1871 and 1878 (Marangoni 1871, 1872, 1878).

These phenomena can be found in the presence of mass transfer and affect liquid–liquid extraction, bubbles, drop migration and the spreading of lubricants. Marangoni motion is also responsible for wall erosion in glass-melting furnaces, and influences crystal growth processes, foam stability, and the stability of liquid layers.

In the particular field of metallurgy, welding procedures are highly affected by surface tension driven motions: the penetration of the liquid phase depends to a great extent on the surface and bulk movements of the liquid pool. The presence of surface active elements, like oxygen and sulphur in liquid iron, changes in a dramatic way the liquid surface tension and its temperature dependence (see Keene 1988), so that the liquid can flow not only with very different velocities but also the direction of flow can be reversed, due to changes in the surface tension coefficient from negative to positive (Desré & Joud 1981).

Thus, considering the importance of surface tension variation as a function of temperature and composition, the high surface activity of oxygen and the scarcity of experimental data on liquid metal systems, we focused our attention on the thermodynamic and kinetic aspects of the oxygen mass transfer at the liquid–vapour interface, relating predictive theoretical models to the experimental results arising from dynamic surface tension measurements of liquid metals by the sessile drop technique.

2. Theoretical approach

Fluid-dynamic models of oxygen mass exchange under very low total pressures have already been developed (Castello *et al.* 1994; Laurant *et al.* 1988) taking into account that, in the Hertz–Knudsen diffusion regime, the mean free path of the molecules in the gas phase is large with respect to the dimensions of the reactor. As a consequence, when metals able to form volatile oxides are considered (see Kellogg 1966), condensation of the oxide vapours on the reactor walls can rapidly occur. In general, any kind of removal of oxide vapours from above the metal surface (pumping, condensation on cold surfaces) can cause significant displacements of the oxidation equilibrium at the gas–liquid interface, by enhancing the rate of the vaporization of the condensed oxide (Brewer & Rosemblatt 1962).

Let us consider a typical experiment of surface tension measurement under a vacuum: the liquid metal sample is placed in a closed chamber at a temperature T, under a total pressure P_{tot} lower than 1 Pa; the oxygen residual pressure $P_{O_2,g}$ is considered constant. Above the oxygen solubility limit in the liquid metal, $x_{O,s}$, the formation of the first stable oxide in the condensed phase is favoured thermodynamically. When the metal and the oxide coexist in the condensed phase, the atmospheric composition is fixed at constant T for the system at equilibrium and it can be calculated from thermochemical data. Moreover, the vapour pressures $P_{j,s}$ of the species containing oxygen resulting from thermal decomposition of the oxide depend exclusively on T. Under these conditions, the oxygen mass transfer between the gas and the condensed phase can be formalized through the evaluation of the global flux to and from the surface. There are two contributions to the oxygen total flux: one given by 'free' molecular oxygen and the other one due to the j species, where oxygen is 'linked' as a chemical compound.

It can be demonstrated (Castello *et al.* 1994) that a steady-state condition with respect to the oxygen exchange is reached when the oxygen residual pressure is equal to the quantity

$$P_{O_2,s}^{E} = P_{O_2,s} + \sum_j v_j \frac{\alpha_j}{\alpha_{O_2}} (m + v_j)^{-1/2} P_{j,s} \tag{2.1}$$

where v_j is the stoichiometry coefficient of the jth oxide formation reaction, $P_{O_2,s}$ is the oxygen equilibrium pressure, $P_{j,s}$ is the vapour pressure of the species containing oxygen, α_i is the sticking coefficient of the species i and m is the ratio of the molecular weights.

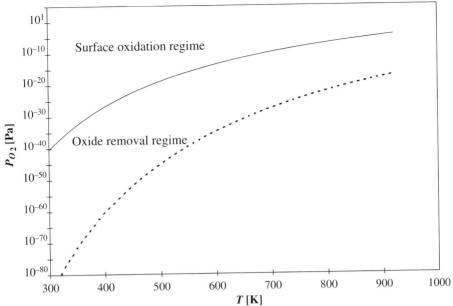

Figure 1. Effective oxidation pressure $P_{O_2,s}^E$ (bold line) and equilibrium oxygen pressure $P_{O_2,s}$ (dotted line) versus T curves for the tin–oxygen system calculated at saturation.

This equation defines the 'effective oxidation pressure' at saturation $(P_{O_2,s}^E)$: it is the oxygen pressure that has to be imposed in order to maintain in a steady state a metal sample completely saturated with oxygen and it applies for the oxidation of liquid and solid metals under vacuum conditions. Owing to the contribution of the volatile oxides to the total oxygen loss, the effective saturation pressure $P_{O_2,s}^E$, can be several orders of magnitude larger than the corresponding saturation value under conditions of thermodynamic equilibrium between the gas and the condensed phase, as shown in figure 1 for the tin–oxygen system.

The $P_{O_2,s}^E$ versus T curve can be regarded as a 'transition curve' between the two regimes: an 'oxygen removal regime' where the metal surface loses oxygen, through the formation of compounds that evaporate rapidly; and, on the other side, a 'surface oxidation regime' where a thermodynamically stable oxide layer is expected to appear on the surface.

An extension of the model briefly described above, allows the interfacial characteristic times (for diffusion, and to reach stationary conditions) and the trend of the interfacial composition versus time to be evaluated. To this end, the theory (Ricci *et al.* 1998) takes into account both the local oxygen mass balance in a liquid metal drop assumed as a rigid sphere under Knudsen conditions, and the total mass balance.

The diffusion time within the drop is

$$t_D = \frac{R^2}{9D_O} \qquad (2.2)$$

and the characteristic time at steady-state conditions is

$$t_{st} = \frac{1}{3} \frac{Rc_l}{N_{O_2,I}^0}(x_O^0 - x_O^{st}), \qquad (2.3)$$

where t_{st} is the time necessary for a homogeneous drop to reach the steady-state

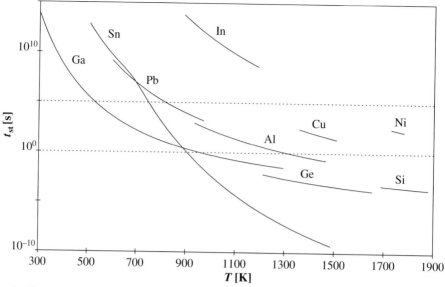

Figure 2. Characteristic times versus temperature for different liquid metal systems between melting and boiling points at a total pressure $P_{\text{tot}} = 1$ Pa. The observable times lie between the two dotted lines.

composition x_O^{st}; R is the drop radius, c_l is the liquid bulk concentration and $N_{O_2,I}^0$ is the initial oxygen flux at the interface.

When the case of nonlinear flux is considered, as in the conditions of the Knudsen regime, and when the saturation values of oxygen pressure and the vapour pressure of species containing oxygen are imposed (oxygen saturation, with the system in monovariant equilibrium, equation (2.1)), the characteristic time can be calculated by the relationship

$$t_{\text{st}} = \frac{R c_l x_{O,s}}{3 k_{O_2} P_{O_2,s}} \left(1 + \sum_j v_j \frac{\alpha_j}{\alpha_{O_2}} \frac{1}{\sqrt{m + v_j}} \frac{P_{j,s}}{P_{O_2,s}} \right)^{-1}, \tag{2.4}$$

where k_{O_2} is the oxygen partial transport coefficient.

The calculated values of the characteristic times as a function of temperature between melting and boiling points for a set of selected metals, considering a metal drop sample of radius $R = 0.5$ cm under a total pressure of 1 Pa, are reported in figure 2. The two dotted lines delimit the time range observable within our experimental set-up (from 1 s to several hours). Among the metals whose curves lie in this range, Sn, Ga and Al seem to be the most convenient to be used as 'test systems'. In the following, the experimental results on liquid tin are presented. Tin has been chosen due to its low melting point and its low vapour pressure, which offers the opportunity to work over a wide temperature range.

3. Experiment

A fast automatic procedure, Automatic Surface-Tension Real-time Acquisition (ASTRA), has been recently designed and setup in our laboratory (Liggieri & Passerone 1989; Ravera *et al.* 1997) allowing the surface tension of liquid metals and alloys to be measured, in a time frame ranging from half a second to hours. In this way, the variation of surface tension with time can be recorded, giving access to the experimental study of dynamic phenomena at liquid surfaces, in particular of adsorption, both from the liquid and the gaseous side.

The experimental setup consists of a vacuum chamber, working both at 10^{-5} Pa and under controlled atmosphere, in which the metallic specimen is introduced by means of a magnetic device; a high quality optical line with a CCD camera; an ion-gun to clean the specimen surface also in the liquid state. The oxygen partial pressure is monitored by means of a solid state oxygen gauge. The metallic specimen is supported on a sapphire cup, especially designed to help maintain the axial symmetry of the sessile drop and to avoid any chemical contamination of the liquid phase.

The experimental reproducibility in measurements of surface tension was found to be ±0.1%, making this the most precise technique for measuring liquid state surface tensions.

Basically, three kinds of experiments have been made, giving access to the dynamic conditions of the liquid tin surface:

(A) at constant temperature, under different oxygen partial pressures, in order to check the calculated value of the effective oxidation pressure $P^{E}_{O_2,s}$;

(B) on 'oxidized' samples, in order to check the effectiveness of the ion-etching procedure to clean the liquid tin surface;

(C) at constant oxygen pressure, but with temperature varying with different gradients, in order to check the response of the surface (through surface tension variations) to the environmental dynamics.

4. Results and discussion

(a) Experiments at constant temperature

The results of two significant experiments, made under essentially the same conditions, i.e. $T = 818$ K and P_{O_2} varying between 10^{-7} and 10^{-1} Pa, are reported here.

Figure 3 shows the results of the variation of the surface tension with time for a liquid tin drop, previously ion-etched ($E = 5$ KeV; $I = 25$ μA) in the liquid state for about 20 min. Figure 4 shows the conditions of oxygen partial pressure and temperature during the experiment.

From the moment of melting, the liquid metal drop under an oxygen partial pressure of 10^{-7} Pa (point 1 in figures 3 and 4) has a surface tension $\gamma = 533$ mN m^{-1}, which remains constant until the imposed oxygen pressure increases to 10^{-1} Pa (point 2 in figures 3 and 4). As shown in figure 4, this variation crosses the $P^{E}_{O_2,s}$ curve, showing that the system goes from the 'oxygen removal' regime to the 'surface oxidation' regime. The surface tension value starts decreasing at the same time, reaching the value $\gamma = 498$ mN m^{-1}. However, when the oxygen pressure is lowered down to 10^{-5} Pa (point 3 in figures 3 and 4), the surface tension no longer varies, even if the representative point lies just under the curve, i.e. in the 'oxide removal' region. Such a behaviour, found also in other experiments, is due to the intrinsic stability of the 'oxide layer' coating the liquid metal surface.

Indeed, if the surface is ion-etched under the same low oxygen pressure, the surface tension immediately increases eventually reaching a value close to the initial one. The results of such a behaviour are reported in figure 5; in figure 6 the corresponding oxygen partial pressure conditions during the experiment are shown. Also in this case, starting from an imposed oxygen partial pressure of 10^{-7} Pa, i.e. the 'oxide

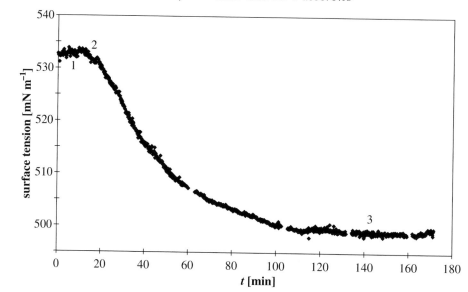

Figure 3. Surface tension versus time plot for liquid tin at $T = 818$ K and varying oxygen partial pressure: 1, $P_{O_2} = 10^{-7}$ Pa; 2, $P_{O_2} = 10^{-1}$ Pa; 3, $P_{O_2} = 10^{-5}$ Pa.

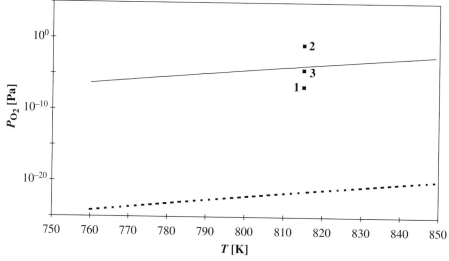

Figure 4. Temperature and oxygen partial pressure conditions of the experiment reported in figure 3: bold line, effective oxidation pressure; dotted line, equilibrium oxygen pressure.

removal' regime (point 1 in figures 5 and 6), the value of the surface tension is $\gamma = 533$ mN m^{-1}, which indicates a 'clean' system.

When the imposed oxygen pressure increases to 10^{-1} Pa (point 2 in figures 5 and 6), this variation crosses the curve, showing that the system goes from the 'oxygen removal' regime to the 'surface oxidation' regime and the surface tension value starts decreasing at the same time. Then, when the oxygen partial pressure is lowered down to 10^{-2} Pa (point 3 in figures 5 and 6), the surface tension continues to decrease but less rapidly than in the previous experiment. However, also in this case, when the oxygen pressure is lowered down to 10^{-5} Pa (point 4 in figures 5 and 6), the surface tension no longer varies, due to the stability of the surface coating

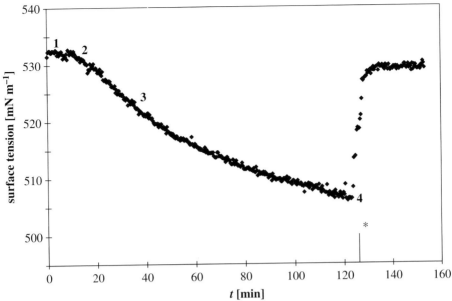

Figure 5. Surface tension versus time plot for liquid tin at $T = 818$ K and varying oxygen partial pressure: 1, $P_{O_2} = 10^{-7}$ Pa; 2, $P_{O_2} = 10^{-1}$ Pa; 3, $P_{O_2} = 10^{-2}$ Pa; 4, $P_{O_2} = 10^{-5}$ Pa; *, start of surface sputtering.

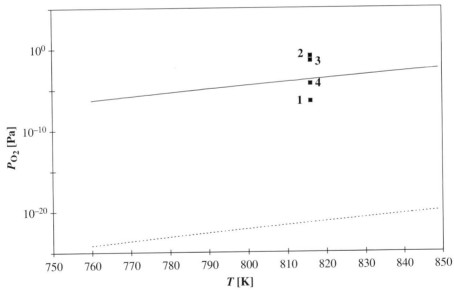

Figure 6. Temperature and oxygen partial pressure conditions of the experiment reported in figure 5: bold line, effective oxidation pressure; dotted line, equilibrium oxygen pressure.

oxide layer. However, an ion-etching of the liquid metal coated surface for about 10 min $(E = 5$ KeV; $I = 25$ μA$)$ is sufficient to increase the surface tension up to a value very close to the initial one. These results confirm the theoretical prediction of an 'effective saturation pressure' well above the thermodynamic one. The calculated oxygen partial pressure values defining the transition between 'oxidized' and 'cleaner' surfaces are in very good agreement with the measured values.

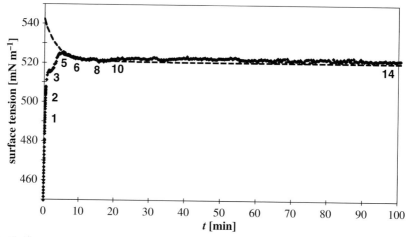

Figure 7. Surface tension versus time plot for liquid tin at $P_{O_2} = 10^{-7}$ Pa and varying temperature; the segment between 10 and 14 is nearly isothermal ($T = 818 \pm 3$ K). Dashed line, reference surface tension values ($\gamma = 560 - 0.125(T - T_m)$).

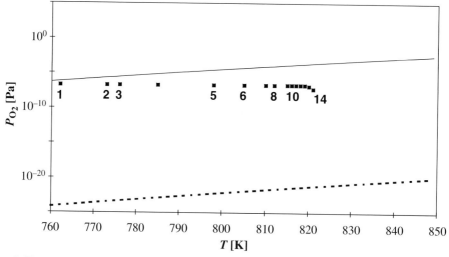

Figure 8. Temperature variation conditions at constant oxygen partial pressure ($P_{O_2} = 10^{-7}$ Pa) of the experiment reported in figure 7: bold line, effective oxidation pressure; dotted line, equilibrium oxygen pressure.

(b) Experiments on 'oxidized' samples

Some measurements have also been made on 'oxidized' samples, to follow the cleaning up of the liquid surface. A typical test is reported in figure 7. Figure 8 shows the corresponding oxygen pressure conditions. The tin sample was pretreated at $T = 823$ K and $P_{O_2} = 90$ Pa for 10 min in order to increase its bulk oxygen content; after quenching, the surface of the sample was cleaned both mechanically and chemically. This sample was then brought under a constant oxygen partial pressure of about 10^{-7} Pa ('oxide removal' regime). It should be noted that the temperature reaches a constant value of 820 K after about 15 min (point 10 in figures 7 and 8). During the first 5 min (from points 1–5 in figures 7 and 8) the value of surface tension rapidly increases corresponding to a decrease of the oxygen content inside the drop.

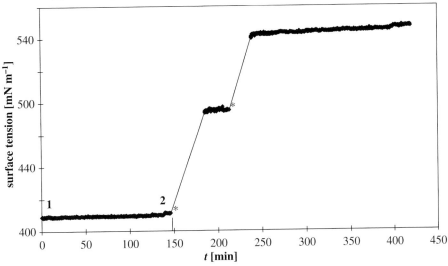

Figure 9. Surface tension versus time plot for liquid tin at $T = 821$ K and varying oxygen partial pressure: 1, $P_{O_2} = 10^{-7}$ Pa; 2, $P_{O_2} = 10^{-1}$ Pa; *, start of surface sputtering.

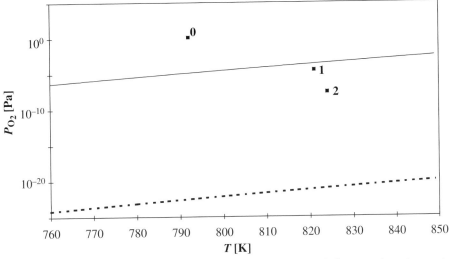

Figure 10. Temperature and oxygen partial pressure conditions of the experiment reported in figure 9. (0 = initial 'oxidized' condition): bold line, effective oxidation pressure; dotted line, equilibrium oxygen pressure.

After this time the experimental data agree very well with the literature (dotted line); the surface tension value remains constant for several hours showing that the clean liquid surface is thermodynamically stable under these conditions.

A further confirmation of the effectiveness of the ion-etching procedure in cleaning the liquid tin surface is shown in figures 9 and 10. Also in this case the sample was put initially under oxidizing conditions starting at $T \approx 790$ K and $P_{O_2} = 1$ Pa for 30 min (point 0 in figure 10) and then brought directly into the 'oxide removal' regime by slightly increasing the temperature (up to $T = 821$ K and $P_{O_2} = 10^{-5}$ Pa: point 1 in figures 9 and 10). When the system enters this region the liquid surface does not react at all: the surface tension remaining constant for nearly 2.5 h. But, as soon

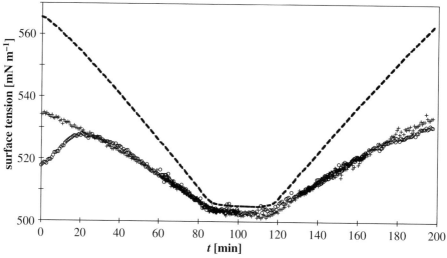

Figure 11. Surface tension versus time variation at 5 K min^{-1} between 625 and 1100 K at different oxygen partial pressures: run 12 ($\circ$), $P_{O_2} = 10^{-3}$ Pa; run 14 (+), $P_{O_2} = 10^{-15}$ Pa; dashed
line, surface tension values ($\gamma = 581 - 0.13(T - T_m)$) from Passerone *et al.* (1990).

as the ion-etching process is started, the surface tension increases very quickly, as
clearly shown in figure 9, and the pressure value decreases spontaneously by three
orders of magnitude (point 2 in figures 9 and 10). The increase in surface tension
seems to be linearly related to the sputtering time.

If the system is left free to evolve after the etching procedure, the surface tension
keeps increasing towards the before mentioned literature value, as expected in the
'oxide removal' regime.

(c) Experiments at constant oxygen partial pressure

The temporal evolution of surface tension of liquid tin as a function of temperature
in the range between 625–1100 K under an imposed constant oxygen partial pressure
has been also investigated. In figure 11, the results of two runs, carried out under
these conditions and at the same temperature variation rate ($\nabla T = 5$ K min^{-1}) are
reported, while the corresponding oxygen partial pressure conditions are reported
in figure 12. The initial conditions of run 12 fall in the 'surface oxidation' regime
below 820 K, while run 14 was carried out in the 'oxide removal' region in the whole
temperature range.

The surface tension results, recorded both during the heating up and the cooling
down periods, show a strikingly linear behaviour as a function of temperature, without any hysteresis effect: the surface tension temperature coefficients coincide, in the
same experiment, during heating and cooling. The different behaviour during the
first 20 min of the two runs depends on the different initial oxygen partial pressure
conditions imposed. From this time on, even if made in the 'surface oxidation' regime
below $T = 820$ K, run 12 has shown surface tension variations which replicate exact-
ly those found in run 14, which instead pertains completely to the 'oxide removal'
regime.

Indeed, under dynamic conditions (very low total pressure, gas flow at low oxygen
partial pressure), the oxygen exchange between the liquid surface and the environment is highly dependent on the rate of temperature variations to which the liquid
surface is subjected. This appears clearly in figure 13 in which the results of run

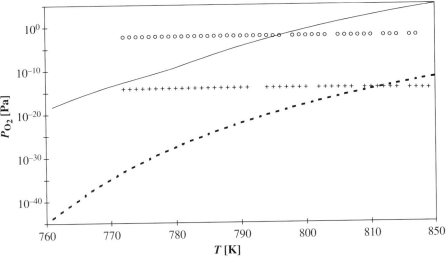

Figure 12. Temperature and oxygen partial pressure conditions of the experiment reported in figure 11: run 12, (○); run 14, (+); bold line, effective oxidation pressure; dotted line, equilibrium oxygen pressure.

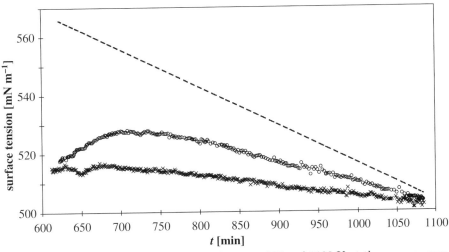

Figure 13. Surface tension versus temperature between 625 and 1100 K at the same oxygen partial pressure, $P_{O_2} = 10^{-1}$ Pa, but different temperature variation rates: run 12 (○), 5 K min^{-1}; run 16 (×), 1 K min^{-1}; dashed line, surface tension values ($\gamma = 581 - 0.13(T - T_m)$) from Passerone *et al.* (1990).

12 and run 16 are compared. They were carried out at the same oxygen partial pressure conditions but at different temperature variation rates, 5 and 1 K min^{-1}, respectively.

It should also be noted that all runs (12, 14 and 16) reach, in the high-temperature limit ($T = 1088$ K), the same high surface tension value ($\gamma = 503$ mN m^{-1}) which is very close to the mean of the reported literature values (Passerone *et al.* 1990). This means that, at high temperature, the liquid surface attains the conditions foreseen by the theoretical model, irrespective of the heating rate; the effect of the temperature gradient becoming relevant at lower temperatures. This behaviour is further

E. Ricci, L. Nanni and A. Passerone

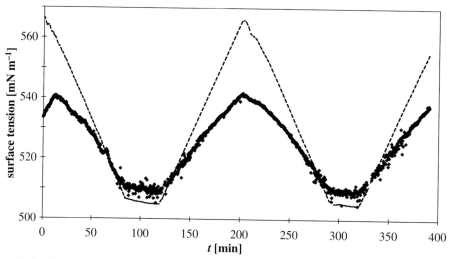

Figure 14. Surface tension versus time variation at 5 K min^{-1} in a double cycle between 625 and 1100 K at constant oxygen partial pressure, $P_{O_2} = 10^{-3}$ Pa: dashed line: surface tension values ($\gamma = 581 - 0.13(T - T_m)$) from Passerone *et al.* (1990).

Table 1. *Surface tension values of tin at melting temperature as a function of imposed oxygen partial pressures and different temperature variation rates*

(In the last column, $T \uparrow$ and $T \downarrow$ denote temperature increasing and temperature decreasing, respectively.)

P_{O_2} (Pa)	∇T (K min^{-1})	γ_0 (mN m^{-1})	γ' (mN m^{-1})	stand. dev. on γ'	remarks	
10^{-3}	5	545.9	-0.073	7.1×10^{-4}	s.ox. $\rightarrow$ ox.rem.	$T \uparrow$
10^{-3}	5	545.6	-0.072	7.8×10^{-4}	ox.rem. $\rightarrow$ s.ox.	$T \downarrow$
10^{-3}	5	554.9	-0.076	5.8×10^{-4}	s.ox. $\rightarrow$ ox.rem.	$T \uparrow$
10^{-3}	5	552.6	-0.073	4.6×10^{-4}	s.ox. $\rightarrow$ ox.rem.	$T \uparrow$
10^{-3}	5	550.6	-0.071	4.7×10^{-4}	ox.rem. $\rightarrow$ s.ox.	$T \downarrow$
10^{-3}	5	552.1	-0.073	5.8×10^{-4}	ox. rem. $\rightarrow$ s.ox.	$T \downarrow$
10^{-3}	1	519.9	-0.037	4.3×10^{-4}	s.ox. $\rightarrow$ ox.rem.	$T \uparrow$
10^{-3}	1	522.6	-0.035	2.5×10^{-4}	s.ox. $\rightarrow$ ox.rem.	$T \uparrow$
10^{-3}	1	527.2	-0.044	5.9×10^{-3}	ox.rem. $\rightarrow$ s.ox.	$T \downarrow$
10^{-15}	5	543.2	-0.067	3.9×10^{-4}	oxide removal	$T \uparrow$
10^{-15}	5	545.5	-0.073	5.8×10^{-4}	oxide removal	$T \downarrow$

confirmed by the experiment shown in figure 14 aimed at confirming the complete reproducibility and reversibility of the surface conditions.

The operating conditions and the results of surface tension measurements of several experiments are collected in table 1. All of them have been made at constant oxygen partial pressure (P_{O_2}) in a temperature range between 625 and 1100 K, and at different temperature variation rates ($\nabla T = 1$ and 5 K min^{-1}).

The remarkable reproducibility of the γ versus $T \uparrow$ and $T \downarrow$ curves within the same class of experiment (1 and 5 K min^{-1}) clearly indicates that the fluid-dynamic characteristics of the oxygen exchange are the key factor in determining the surface tension variations.

5. Conclusions

On the basis of the results presented here the following conclusions can be drawn.

(1) The experimental technique (ASTRA) allows real-time surface tension data of liquid metal systems to be measured in a time-scale from 0.5 s to hours, with a precision of 0.1%. Thus, the dynamic behaviour of the liquid surface in various environments can be easily followed, and possible 'second-order' effects on surface tension can be detected.

(2) The theoretical prediction of an 'effective saturation pressure', usually many orders of magnitude higher than the thermodynamic prediction, is fully confirmed: this new limit can help explain the oxidation behaviour of liquid metals under a vacuum. Moreover, a precise definition of the operating conditions to achieve a clean surface can be given.

(3) The behaviour of liquid tin (and possibly of other liquid metals) with respect to surface oxidation depends, at constant temperature, on the fluid-dynamic regimes of metal and oxide vapours exchange (as monitored by the $P_{O_2,s}^E$ critical curve) and, at varying temperature, on its gradient with time.

(4) As Marangoni effects are driven by the surface tension variations with temperature and/or composition, special care has to be taken when using literature values. Indeed, the present study clearly shows that even if 'equilibrium' positive values are expected for γ' it may happen that, under dynamic conditions, these values change, not only in magnitude, but also in sign. From a technological viewpoint, the findings reported here imply that Marangoni motions are affected not only by temperature gradients but also, and to a large extent, by the rate imposed to temperature variations.

The contribution to the theoretical development of this work by Professor P. Costa and Professor E. Arato (ISTIC, University of Genova) is gratefully acknowledged.

References

Brewer, L. & Rosemblatt, G. M. 1962 *Trans. Met. Soc. AIME* **224**, 1268.

Castello, P., Ricci, E., Passerone, A. & Costa, P. 1994 *J. Mater. Sci.* **29**, 6104.

Desré, P. J. & Joud, J. C. 1981 *Acta Astronaut.* **8**, 407.

Keene, B. J. 1988 *Int. Mater. Rev.* **33**, 1.

Kellogg, H. H. 1966 *Trans. Met. Soc. AIME* **263**, 602.

Laurant, V., Chatain, D., Chatillion, C. & Eustathopoulos, N. 1988 *Acta Metall.* **36**, 1797.

Liggieri, L. & Passerone, A. 1989 *High Temp. Tech.* **7**, 80.

Marangoni, C. 1871 *A. Phys. Chem., Poggendorff* **143**, 337.

Marangoni, C. 1872 *Nuovo Cim.* ser. 2, **5/6**, 239.

Marangoni, C. 1878 *Nuovo Cim.* ser. 3, **3**, 50, 97, and 193.

Passerone, A., Ricci, E. & Sangiorgi, R. 1990 *J. Mater. Sci.* **25**, 4266.

Ravera, F., Liggieri, L., Ferrari, M. & Passerone, A. 1997 *IXth Int. Conf. on Surface and Colloids Sci. (IACIS), Sophia.*

Ricci, E., Nanni, L., Arato, E. & Costa, P. 1998 *J. Mater. Sci.* **33**, 305.

Discussion

J. JEFFES (*Imperial College, London, UK*). D. Richardson found the rate of oxygen transfer from gas to a levitated copper drop was very sensitive to surface contamination. Very small amounts of silicon formed a permanent layer of silica. This was overcome by solidifying the droplet while levitated, treating it with hydrofluoric acid and relevitated when the oxygen mass transfer was found to be ten times greater.

In the present experiments was it really proved that the tin surface was really clean? Was oxygen content of the surface determined?

E. RICCI. No surface analysis was performed on the molten tin drops.

However, experiments have been made where the liquid surface was ion-etched until the surface tension value reached the high value expected for 'pure' tin. The reproducibility and the reversibility of the results obtained in that way represents an indirect confirmation that the liquid surface was reasonably clean when its surface tension attained the highest values.

Influence of reactive solute transport on spreading kinetics of alloy droplets on ceramic surfaces

By N. Eustathopoulos[1], J. P. Garandet[2] and B. Drevet[2]

[1]Laboratoire de Thermodynamique et Physico-Chimie Métallurgiques, UMR 5614 CNRS-INPG-UJF, Ecole Nationale Supérieure d'Electrochimie et d'Electrométallurgie de Grenoble, Institut National Polytechnique de Grenoble, BP 75 Domaine Universitaire, 38402 Saint-Martin d'Hères Cedex, France
[2]Commissariat à l'Energie Atomique, DTA/CEREM/DEM/SPCM, Laboratoire de Recherche de Base en Solidification, 17 rue des Martyrs, 38054 Grenoble Cedex 9, France

This paper focuses on the question of the processes which can be rate-limiting for reactive spreading in the sessile drop configuration. It will be shown that for a class of systems, spreading kinetics is controlled by the transport of species involved in the reaction between the drop bulk and the triple line. For these systems convection, and especially Marangoni convection, may significantly affect the dynamics of wetting.

Keywords: wetting; reactivity; interfaces; diffusion; Marangoni convection

---

1. Introduction

In non-reactive metal–ceramic systems non-wetting is usually observed, the angle θ formed at the contact line of three phases, solid (S), liquid (L) and vapour (V), being greater than 90° (Naidich 1981; Eustathopoulos & Drevet 1994). Typical examples are the couples $Cu-Al_2O_3$ and $Cu-C_{gr}$, for which the contact angle under high vacuum or in inert gas is as large as 130–140° (Naidich 1981). However, in several fields of materials science, e.g. joining of ceramics by brazing alloys or metal–ceramic multimaterial processing by infiltration routes, good wetting is required (θ much lower than 90°). Although some improvement in wetting and adhesion may be produced by tensioactive solutes (Eustathopoulos & Drevet 1994), acting by adsorption at the metal–ceramic interface, strong effects on these properties can be obtained by specific reactive solutes forming, by reaction with the ceramic at the interface, continuous layers of a new compound (Nicholas 1986; Loehman & Tomsia 1988; Nogi 1993). Due to uncertainties on the driving force of reactive wetting and to the complexity of kinetic phenomena in the sessile drop configuration, no fundamental approach has been developed until recently on spreading kinetics in reactive systems.

This paper reviews the main results obtained on the dynamics of wetting in reactive systems during the period 1993–96. In §2 the question of reactive wetting driving force is discussed. In §3, the different processes which may be rate-limiting are considered, namely viscous flow, local chemical kinetics and solute diffusion. A new analysis of the possible effect on wetting kinetics of convection in the liquid drop

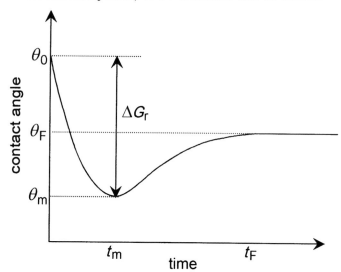

Figure 1. Contact angle versus time curve in a reactive system according to the model of Aksay *et al.* (1974).

is also presented. All experimental results given below were obtained by the sessile drop technique under high vacuum or in inert gas, using millimetre-size droplets and smooth (average roughness of a few nm) monocrystalline (α-Al$_2$O$_3$, SiC) or vitreous (carbon) substrates. Although different kinds of reaction can affect wetting (for instance the simple dissolution of an oxide in the liquid alloy (Eustathopoulos & Drevet 1994)) the paper focuses on interfacial reactions leading to the formation of a dense layer of solid reaction product.

2. Driving force of reactive wetting

When a pure liquid wets the smooth and chemically homogeneous surface of an inert solid, the wetting driving force at time t is given by

$$F_{\mathrm{d}}(t) = \sigma_{\mathrm{SV}}^0 - \sigma_{\mathrm{SL}}^0 - \sigma_{\mathrm{LV}}^0 \cos \theta(t), \tag{2.1}$$

where σ_{ij}^0 are the characteristic surface energies of the system, and $\theta(t)$ is the instantaneous contact angle. At equilibrium, $F_{\mathrm{d}} = 0$, which leads to the classical Young's equation as follows:

$$\cos \theta^0 = \frac{\sigma_{\mathrm{SV}}^0 - \sigma_{\mathrm{SL}}^0}{\sigma_{\mathrm{LV}}^0}. \tag{2.2}$$

For reactive solid–liquid systems, no clear definition of the driving force exists at the present time. Aksay *et al.* (1974) replaced the σ_{SL}^0 term in equation (2.1) by

$$\sigma_{\mathrm{SL}}(t) = \sigma_{\mathrm{SL}}^0 + \Delta G_{\mathrm{r}}(t), \tag{2.3}$$

where $\Delta G_{\mathrm{r}}(t)$ is the change in Gibbs energy released per unit area by the reaction in the 'immediate vicinity of the solid–liquid interface' (Aksay *et al.* 1974).

Aksay *et al.* (1974) argue that the effect of the $\Delta G_{\mathrm{r}}(t)$ term is strongest during the early stages of contact because the interfacial rate is at its maximum when the liquid contacts a fresh unreacted solid surface. Thereafter, the reaction kinetics slow down, and after an initial decrease, the contact angle increases and gradually approaches the equilibrium value (figure 1).

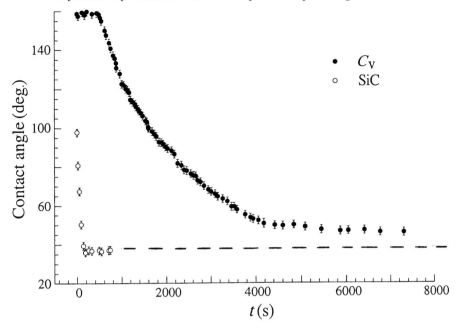

Figure 2. Contact angle kinetics obtained for a Cu-40 at.% Si alloy on vitreous carbon and
α-silicon carbide substrates under high vacuum at 1423 K (Landry *et al.* 1997).

A critical discussion of this model, as well as of experiments used to validate it, is given elsewhere (Eustathopoulos 1996). In the present paper, we will describe two recent experiments which disagree with Aksay's model and allow us to propose another interpretation of reactive driving force. The first experiment is a Cu–Si alloy on vitreous carbon (figure 2) (Landry *et al.* 1997). Pure copper does not wet vitreous carbon (at 1150 °C, $\theta = 137 \pm 5°$) but a Cu–40 at.% Si alloy wets well this substrate due to the formation, at the interface, of a continuous submicron layer of SiC. When the experiment is repeated with the same alloy on an α-SiC monocrystalline substrate wetting kinetics are very different: very fast in the non-reactive system (i.e. Cu–Si/SiC) and very slow in the reactive one (Cu–Si/Cv). In fact in this system the wettable silicon carbide 'substrate' is fabricated *in situ* and this process takes a certain time. The curves of figure 2 show first that the steady contact angles in the reactive and the non-reactive systems are nearly the same, and second that the contact angle decreases monotonically with time to its steady value, in disagreement with Aksay's model.

A slightly different procedure was used in another experiment performed with a Cu-1 at.% Cr alloy on a Cv substrate (Landry *et al.* 1997). Chromium promotes wetting of copper, forming at the interface a continuous layer a few microns in thickness of the wettable metallic-like chromium carbide Cr_7C_3. After cooling, the Cu–Cr solidified drop was dissolved, a small quantity of a Cu-1 at.% Cr alloy was placed in the centre of the carbide layer and the wetting experiment was repeated at the same temperature. Results (figure 3) are very similar to that of the Cu–Si alloy with respect to the following three points: (i) spreading in the reactive couple is much slower than in the corresponding non-reactive couple; (ii) the final contact angle formed by the Cu–Cr alloy on the initial substrate (C_v) and on the reaction product (chromium carbide) are nearly the same; and (iii) in the reactive system no minimum of θ is observed.

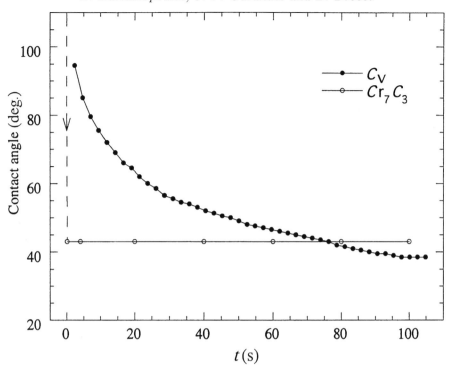

Figure 3. Contact angle versus time curves for a Cu-1 at.% Cr alloy on vitreous carbon and
Cr$_7$C$_3$ substrates at 1373 K (Landry *et al.* 1997).

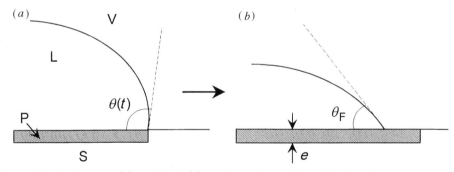

Figure 4. Instantaneous (*a*) and final (*b*) configuration at the solid–liquid–vapour triple line
during spreading in reactive wetting.

From these and other experiments (Espié *et al.* 1994; Kritsalis *et al.* 1994), it was
concluded that wetting in reactive systems correlates with the final interfacial chemistry at the triple line, not with the intensity of interfacial reactions (Eustathopoulos
& Drevet 1993, 1994). Therefore, the reactive wetting driving force is

$$F_\mathrm{d}(t) = \sigma_\mathrm{LV}^0[\cos\theta_\mathrm{F} - \cos\theta(t)], \qquad (2.4)$$

where θ_F is the equilibrium contact angle of the liquid on the reaction product surface
(figure 4*b*) (note that this driving force is equal to the driving force in non-reactive-
wetting of a liquid drop on an infinite surface of reaction product). The question
discussed below is what are the processes which limit the velocity of the triple line
when the contact angle changes from the initial contact angle to θ_F.

3. Spreading kinetics

In non-reactive systems, the spreading rate is controlled by the viscous flow and described (for $\theta < 60°$) by a power function of drop base radius R versus time t:

$$R^n \approx t, \qquad (3.1)$$

in which n, as calculated by Tanner (1979) and by de Gennes (1985) is equal to 10.

Because the viscosity of molten metals is very low, the time needed for millimetre size droplets to reach capillary equilibrium is less than 10^{-1} s (Naidich 1981; Eremenko *et al.* 1994; Laurent 1988). This time is several orders of magnitude shorter than the spreading times observed in reactive metal-ceramic systems, usually lying in the range 10^1–10^4 s (Naidich 1981; Loehman & Tomsia 1988; Espié *et al.* 1994; Landry *et al.* 1997; Fujii *et al.* 1993; Nicholas & Peteves 1994; Mortimer & Nicholas 1970, 1973). Therefore in the latter systems the rate of spreading is not controlled by viscous resistance, but by the interfacial reaction itself. Another consequence of the low viscosity of the liquid is that during reactive spreading the drop radius is equal to the reaction product radius, i.e. the positions of the triple line and of the radial reaction front are identical.

Hereafter, spreading kinetics will be discussed by means of the classical two-step scheme used in treating kinetic phenomena, consisting of a local process at the interface and transport phenomena in bulk materials. The only difference in reactive wetting, but an important one, is that the relevant defect is not a two-dimensional interface but a line defect, the contact line of three phases S, L and V: because the liquid has a direct access to the solid at the triple line, the reaction rate at this particular point is two to three orders of magnitude higher than the reaction rate at the interface far from the triple line where the reaction occurs by slow diffusion through a solid layer (Landry & Eustathopoulos 1996).

In the framework of this general description, two limiting cases can be defined depending on the rate of the chemical reaction at the triple line compared to the rate of transport of reactive solute from the drop bulk to the triple line (or of a soluble reaction product from the triple line to the drop bulk).

(a) Reaction-limited spreading

In the first limit case chemical kinetics at the triple line are rate-limiting because transport within the droplet is comparatively rapid (or not needed when the drop is made of a pure reactive metal). In this case, (i) if the reaction does not change the global drop composition significantly, which means that the chemical environment of the triple line is constant with time and (ii) if a steady configuration is established at the triple line during wetting, then the rate of reaction and hence the triple line velocity are constant with time (Landry & Eustathopoulos 1996):

$$R - R_0 = Kt, \qquad (3.2)$$

where R_0 is the initial drop base radius and K is a system constant, independent of the drop volume V_d.

An example is pure Al on vitreous carbon under high vacuum (figure 5) (Landry & Eustathopoulos 1996). In this system, wetting is promoted by the formation of a continuous layer of micron thickness of the wettable aluminium carbide at the interface, the final contact angle θ_F being about 70° (figure 5). After an initial stage, between time $t = 0$ and time $t_1 = 400$ s, due to deoxidation of the Al drop, the

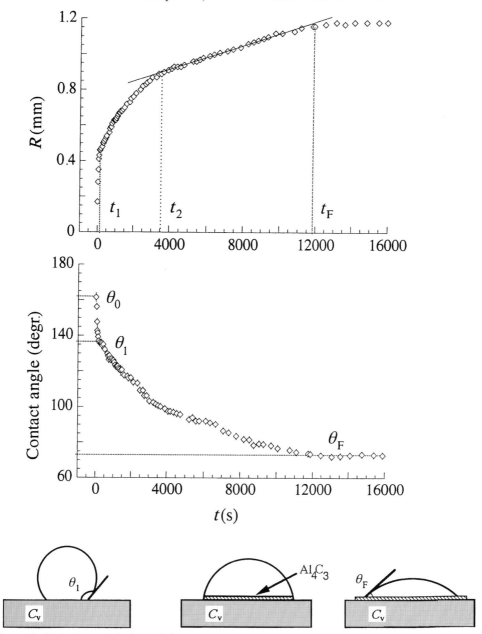

Figure 5. Variations with time of the contact angle and drop base radius observed in the Al/C_v system at 1100 K (Landry & Eustathopoulos 1996) and schematic representation of interfaces at $t = t_1$ (left), $t_2 < t < t_F$ (middle) and $t > t_F$ (right).

spreading curve $R(t)$ presents a nonlinear part (from t_1 to t_2) followed by a linear part (from t_2 to t_F).

The contact angle θ_1 is the contact angle of pure deoxidized Al on the original unreacted C_v surface (figure 5, bottom). θ_2 is the first contact angle corresponding to an interface fully covered by a reaction product layer. Therefore, the decrease from θ_1 to θ_2 corresponds to a transition from a non-reacted to a completely reacted interface. For $t > t_2$, a steady configuration is established at the triple line and, as a result, the reaction rate and the triple line velocity are constant with time. During this stage the macroscopically observed contact angle is not related to capillary force equilibrium but is dictated by the drop volume and the radius of the reaction product layer. Indeed, the advance of the liquid is hindered by the presence of a non-wettable vitreous carbon in front of the triple line. Thus, the only way to move ahead is by lateral growth of the wettable carbide layer until the macroscopic contact angle equals the equilibrium contact angle of Al on the carbide.

This interpretation does not take into account the possible effect of a reaction occuring ahead of the triple line. In vacuum, such a reaction may occur by evaporation–condensation. For geometrical reasons (evidenced by considering the direction of evaporation with regard to the substrate surface) this mechanism is effective for non-wetting drops ($\theta > 90°$) and therefore may be, at least partially, responsible of the rapid spreading observed between t_1 and t_2 (figure 5) (Dezellus *et al.* 1998).

'Linear wetting' can occur in different systems and for different types of reaction. Examples are Cu–Si alloys on oxidized SiC (Rado 1997), the reactive Cu–Ag–Ti/Al$_2$O$_3$ system (Nicholas & Peteves 1995) or Cu–Si alloys on Cv (Landry *et al.* 1996; Dezellus *et al.* 1998). In some cases, small but significant deviations from linearity were observed and attributed to roughness of the reaction layer delaying the movement of the triple line by pinning (Landry & Eustathopoulos 1996).

(b) Transport-limited spreading

When local reaction rates are comparatively high, the rate of lateral growth of the reaction product at the triple line is limited by the supply of reactant from the drop bulk to the triple line. Because the contact angle decreases continuously during wetting, the reduction in transport field will lead to a continuous decrease in the reaction rate and, as a result, of the rate of movement of the triple line itself (figure 4). Therefore, time-dependent spreading rates are expected in this case (Landry & Eustathopoulos 1996).

In the liquid, solute is transported by convection and diffusion and the governing equation is Fick's second law written in the referential of the triple line moving with a velocity V_{TL} ($V_{\mathrm{TL}} = \mathrm{d}R/\mathrm{d}t$):

$$\frac{\partial C}{\partial t} + V_{\mathrm{F}} \cdot \nabla C = D\nabla^2 C + V_{\mathrm{TL}} \cdot \nabla C. \tag{3.3}$$

In this equation, the concentration C of reactive species is expressed in mass fraction. V_{F} denotes the local velocity of the fluid and results both from the movement of the triple line and from convection generated by temperature and concentration gradients in the liquid.

(i) Pure diffusion

Given the complexity of the real situation in the sessile drop configuration, a simplified analysis of diffusion-limited spreading has been proposed (Mortensen *et al.* 1997), in which it was assumed that in a small volume near the triple line diffusion is the dominating mechanism for solute transport. Inside this volume modelled as a straight wedge of angle θ (figure 6), the velocity V_{F} is taken equal to V_{TL} such that equation (3.3) reduces to

$$\frac{\partial C}{\partial t} = D\nabla^2 C. \tag{3.4}$$

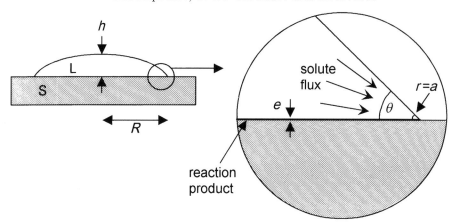

Figure 6. Schematic description of the advancing triple line driven by localized chemical reaction requiring solute transport in the liquid.

Neglecting the reaction at the interface far from the triple line, equation (3.4) was solved in cylindrical coordinates (r, Φ) (Mortensen *et al.* 1997). The main result of this analysis is that, although diffusion in the sessile drop configuration is basically unsteady, an accurate steady-state solution was found in which the reaction rate at the triple line, and hence the spreading rate, depends only on the configuration of the triple line or, in other words, on the instantaneous contact angle θ (Mortensen *et al.* 1997):

$$\frac{\mathrm{d}R}{\mathrm{d}t} = \frac{2DF(C_0 - C_e)\theta}{en_v}. \tag{3.5}$$

In this equation, F is a constant close to 0.04, e is the thickness of the reaction product layer at the triple line and n_v is the number of moles of reactive species per unit volume of the reaction product. C_0 is the concentration of the reactive species in the bulk drop and C_e is the equilibrium concentration assumed to be attained inside a small volume of radius a near the triple line (see figure 6) (a being on the order of a few atomic jumps in the liquid, typically 10^{-9} m). Note that for the sake of homogeneity, the units of C_0 and C_e in equation (3.5) are moles per unit volume of the liquid. For millimetre-size droplets, forming nearly spherical cups, and contact angles not too high ($\theta \leqslant 60°$), θ is closely approximated by $4V_d/\pi R^3$ where V_d is the drop volume. Equation (3.5) is then easily integrated to give

$$R^4 - R_0^4 = Kt, \tag{3.6}$$

where K is a constant for a given system, temperature, drop volume and concentration C_0.

In the framework of this model, solute gradients are localized near the triple line within a volume of radius δ approximately equal to $a \exp(1/(2F))$. For values of a of a few nm, δ is of the order of tens of micrometers. This discussion allows to check *a posteriori* the validity of the quasi-stationary approximation, using an approach similar to that followed in the related problem of mass transport ahead of a solidification interface (Garandet 1993). Indeed, let us denote τ the typical time scale of the composition field; the time-dependent and the Laplacian term in equation (3.4) will thus be proportional to $(1/\tau)$ and (D/δ^2), respectively. In our present problem, taking $\tau = 100$ s, $D = 3 \times 10^{-9}$ m^2 s^{-1} and $\delta = 20$ μm, $D\tau/\delta^2$ is of the order of 1000 and the process can thus be safely taken as quasi-stationary.

Existing data on reactive wetting in metal–ceramic systems are usually given only as plots of contact angle versus time. Although from these results it may be concluded that in many systems spreading is nonlinear, no quantitative comparison between these data and equation (3.6) is possible. In a recent study, for the high reaction rate Cu-1 at.% Cr/C$_v$ system, for which the spreading time is only 100 s, i.e. about 100 times shorter than in Al/C$_v$ couple, the value of exponent n giving the best fit of the experimental $R(t)$ curves was found to be in the range 5.5–6.5 (Voitovich *et al.* 1998). These values are very different from $n = 10$, which is characteristic of viscous spreading, and from $n = 1$, typical of reaction-limited spreading. However, the values are higher than $n = 4$ calculated for pure diffusion. A possible reason for this disagreement is that convection in the drop has been neglected in this model.

(ii) *Convection*

In experimental practice, one can expect that solute will also be carried out by convective movements in the drop. Indeed, the density gradients related to temperature and composition differences interact with gravity to sustain a fluid motion. In addition, temperature and composition differences at the liquid–vapour interface give rise to Marangoni convection. To account for this convective effect, we shall rely on a former work carried out in a crystal growth configuration, where an equation similar to equation (3.3) governs solute repartition in the melt. It was seen in Garandet (1993) that both advection and convection could be combined in a single 'effective' velocity V_{eff}, defined as

$$V_{eff} = D/\delta = V_{TL} - V_F(\delta), \tag{3.7}$$

where δ stands for the solutal boundary layer thickness. Note that with our choice of notation $V_F(\delta)$, the fluid flow velocity at the boundary layer scale, is negative, but that V_{eff} is always positive (Garandet 1993).

In this section, we shall suppose that convection is a relevant mass transport mechanism, i.e. that fluid flow carries a significant amount of solute to the triple line. Whatever the physical origin of fluid flow, its expected effect will be to increase solute transport and to reduce the thickness δ of the solutal boundary layer in the vicinity of the triple line. In the mathematical formulation of the problem, we can thus safely neglect the time dependence of the composition field and consider the process as quasi-stationary, as discussed above. We thus look for a solution to the equation

$$D\nabla^2 C + V_{eff} \cdot \nabla C = 0, \tag{3.8}$$

C being the solute composition expressed in mass fraction. At this point, V_{eff} should be taken as a mathematical auxiliary, but we shall return later to its physical basis.

In the frame of the present work, we are more interested in the identification of relevant transport mechanisms than with an accurate description of the problem geometry. We shall thus simplify matters and use a one-dimensional approach in Cartesian coordinates (see figure 7) to derive the composition gradient in the vicinity of the triple line. Equation (3.8) thus becomes

$$D\mathrm{d}^2 C/\mathrm{d}x^2 + V_{eff}\mathrm{d}C/\mathrm{d}x = 0. \tag{3.9}$$

The solution to equation (3.9), along with boundary conditions $C = C_e$ in $x = a$ and $C = C_0$ as $x \to \infty$ is simply

$$C = \exp(V_{eff}a/D)(C_e - C_0)\exp(-V_{eff}x/D) + C_0. \tag{3.10}$$

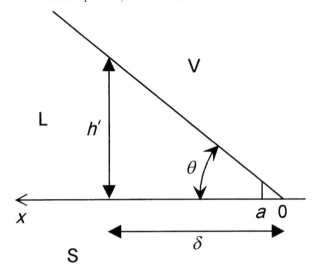

Figure 7. One-dimensional representation of the vicinity of the triple line at the solutal boundary layer scale δ.

Using this expression, the mass flux at the abscissa $x = a$ can be written as

$$|J| = \rho_{\mathrm{L}} D \frac{\mathrm{d}C}{\mathrm{d}x}\Big|_{x=a} = \rho_{\mathrm{L}}(C_0 - C_{\mathrm{e}})V_{\mathrm{eff}}, \qquad (3.11)$$

where ρ_{L} is the mass density of the liquid. Assuming, that θ is small enough to ensure $\tan\theta \approx \theta$, the integrated mass flux at the abscissa $x = a$ is given by

$$|J_{\mathrm{t}}| = \rho_{\mathrm{L}}(C_0 - C_{\mathrm{e}})V_{\mathrm{eff}}2\pi R\theta a. \qquad (3.12)$$

In this expression of the total mass flux, we implicitly supposed convective transport to be negligible at the scale a, which is a fairly safe assumption. The mass balance in $x = a$ can be written as

$$\rho_{\mathrm{L}}(C_0 - C_{\mathrm{e}})V_{\mathrm{eff}}\theta a = \rho_{\mathrm{P}}C_{\mathrm{P}}e(\mathrm{d}R/\mathrm{d}t). \qquad (3.13)$$

In the above expression ρ_{P} and C_{P} denote the mass density and the solute mass fraction of the product, respectively.

We now have to specify in more detail the physical sources of fluid flow in order to obtain quantitative informations about the spreading kinetics. Our first hypothesis will be to consider that solutal convection is negligible with respect to thermal convection. This may seem to be an *a priori* surprising assumption since it is well known that density and surface tension can both be very sensitive to the presence of solutes, even at low concentrations in the case of tensioactive species. However, we have seen that the composition variation, as evaluated by the pure diffusion model, occurs on a very limited length scale (*ca.* $\delta = 20\,\mu\mathrm{m}$), meaning that the flow has very little room to develop. Indeed, an order of magnitude analysis carried out in a recent work (Alboussiere *et al.* 1997) indicates that an *a priori* higher density gradient due to composition differences is less efficient as a convective driving force than temperature-induced density variations, the reason being that the latter take place over the whole fluid body. In the present paper, we shall only consider thermal convection but it should be kept in mind that in concentrated systems with large variations of density with composition, or in the case of tensioactive species, solutal convection could have an impact.

The next step is to estimate the relative influences of bulk and Marangoni convection in a spreading drop. To do so, we shall rely on the parallel flow solution obtained by Birikh (1966) in a layer of height h submitted to a constant thermal gradient, where the velocity at the liquid–vapour interface is given by

$$U = (\nu/h)[-\tfrac{1}{4} Re_{\mathrm{M}} + \tfrac{1}{48} Gr], \tag{3.14}$$

where ν represents the kinematic viscosity of the fluid. The non-dimensional Reynolds–Marangoni (Re_{M}) and Grashof (Gr) numbers are defined as

$$Re_{\mathrm{M}} = \sigma'_{\mathrm{T}} G h^2/\rho_{\mathrm{L}}\nu^2, \tag{3.15}$$

$$Gr = \beta_{\mathrm{T}} g G h^4/\nu^2. \tag{3.16}$$

The above expressions include thermophysical properties of the fluid, namely ρ_{L}, ν, σ'_{T}, derivative of its surface tension with respect to temperature and β_{T} its thermal expansion coefficient. The experimental conditions are characterized by h, the drop height, and G, the thermal gradient in the drop, with g denoting the acceleration due to gravity. Taking reasonable values, e.g. $\rho_{\mathrm{L}} = 10 \ \mathrm{kg \ m^{-3}}$, $\nu = 3 \times 10^{-7} \ \mathrm{m^2 \ s^{-1}}$, $\sigma'_{\mathrm{T}} = 2 \times 10^{-4} \ \mathrm{J \ m^{-2} \ K^{-1}}$, $\beta_{\mathrm{T}} = 10^{-4} \ \mathrm{K^{-1}}$, $g = 10 \ \mathrm{m \ s^{-2}}$ and $h = 1 \ \mathrm{mm}$, we get

$$Re_{\mathrm{M}} = 0.22G, \quad Gr = 0.01G \quad (G \ \mathrm{in \ K \ m^{-1}}). \tag{3.17}$$

It thus appears clearly from equation (3.14) that bulk natural convection can be safely neglected in comparison with Marangoni convection. To quantify matters, we need to estimate the thermal gradient inside the drop. In typical sessile drop isothermal furnaces, residual thermal gradients lie between 0.1 and $0.5 \ \mathrm{K \ mm^{-1}}$. Taking $Re_{\mathrm{M}} = 100$, we find that the fluid velocity in the drop is in the $\mathrm{cm \ s^{-1}}$ range. Incidentally, at such a limited value of Re_{M}, the flow is expected to be laminar.

Nevertheless, we still have to estimate the magnitude of the fluid velocity V_{F} at the composition boundary layer scale δ. As discussed in the related problem of solutal transport in crystal growth configurations (Garandet 1993), this is indeed the scale at which convection needs to be effective. Since we deal here with thin boundary layers, we should expect V_{F} to be significantly smaller than U. To get a rough estimate for $V_{\mathrm{F}}(\delta)$, we assume that the balance between surface tension and viscous forces used to derive equation (3.14) can be written at the boundary layer scale, i.e. when the drop height is h'. As can be seen in figure 7, h' and δ are related by $\tan\theta = h'/\delta$, meaning that they are of the same order of magnitude. The fluid velocity at the scale δ can thus be estimated as

$$V_{\mathrm{F}}(\delta) = (|\sigma'_{\mathrm{T}}|G/\rho_{\mathrm{L}}\nu)\tfrac{1}{4}h'. \tag{3.18}$$

Taking $h' = 40 \ \mu\mathrm{m}$ (resp. $h' = 10 \ \mu\mathrm{m}$) along with the above values and $G = 0.5 \ \mathrm{K \ mm^{-1}}$ we get $V_{\mathrm{C}} = 300 \ \mu\mathrm{m \ s^{-1}}$ (resp. $V_{\mathrm{C}} = 75 \ \mu\mathrm{m \ s^{-1}}$). Assuming once more that θ is small enough to ensure $\tan\theta \approx \theta$, we get

$$V_{\mathrm{F}}(\delta) = (|\sigma'_{\mathrm{T}}|G/\rho_{\mathrm{L}}\nu)(\tfrac{1}{4}\delta\theta). \tag{3.19}$$

The above values of the convection velocity are indeed much smaller than U, but they are still significantly higher than the velocity of the triple line which is of the order of $10 \ \mu\mathrm{m \ s^{-1}}$ in many practical cases (Voitovich *et al.* 1998; Drevet *et al.* 1996). We can thus assume that convection is indeed the dominant solute transport mode and drop the V_{TL} term in equation (3.7) or in other words set $V_{\mathrm{eff}} = -V_{\mathrm{F}}(\delta)$. The relation $V_{\mathrm{eff}} = D/\delta$ thus yields

$$\delta = 2(\rho_{\mathrm{L}}\nu D/|\sigma'_{\mathrm{T}}|G)^{1/2}\theta^{-1/2}. \tag{3.20}$$

Reporting this value of δ in equation (3.19) and in turn equation (3.19) in the mass balance equation (3.13), we get

$$(1/\theta^{3/2})\mathrm{d}R/\mathrm{d}t = \tfrac{1}{2}(\rho_\mathrm{L}/\rho_\mathrm{P})[(C_0 - C_\mathrm{e})/C_\mathrm{P}](a/e)(|\sigma'_\mathrm{T}|GD/\rho_\mathrm{L}\nu)^{1/2}. \qquad (3.21)$$

Assuming once more that $\theta \ll 90°$ and setting $\theta = 4V_\mathrm{d}/\pi R^3$, we find that $R^{9/2}\,\mathrm{d}R/\mathrm{d}t$ is constant, and we finally get upon integration

$$R^{11/2} - R_0^{11/2} = \alpha t, \qquad (3.22)$$

where α represents a proportionality factor given by

$$\alpha = 22/\pi^{3/2}(\rho_\mathrm{L}/\rho_\mathrm{P})[(C_0 - C_\mathrm{e})/C_\mathrm{P}](a/e)(|\sigma'_\mathrm{T}|GD/\rho_\mathrm{L}\nu)^{1/2}V_\mathrm{d}^{3/2}. \qquad (3.23)$$

We have thus obtained an analytical expression for the drop spreading kinetics when temperature-driven Marangoni convection can be considered as the dominant solutal transport mechanism. Experimental results for the Cu-1 at.% Cr/C$_\mathrm{v}$ system ($n = 5.5$–6.5) (Voitovich *et al.* 1998) agree much better with equation (3.22) than with equation (3.6) established in the case of diffusive transport. Such a good agreement between experiment and the model based on Marangoni convection should be taken as coincidental in view of the large number of simplifying assumptions. The most significant result of this analysis is that in the case of millimetre-size droplets, Marangoni convection can significantly increase solutal transport with regard to the purely diffusive regime.

4. Conclusions

From the analysis of experimental data for model metal–ceramic systems, it appears that the final contact angle in a reactive system is given with good accuracy by Young's (or the equilibrium) contact angle of the liquid on the reaction product. Reactive systems feature either linear or nonlinear $R(t)$ spreading, corresponding to reaction-controlled and to transport-controlled regimes, respectively.

One of the purposes of the present study was to discuss the possible effects on spreading kinetics of various types of convection, i.e. thermal/solutal and bulk/Marangoni. Except for reactive solutes, which are also tensioactive at the liquid alloy free surface, or for alloys highly concentrated in reactive solutes, Marangoni convection generated by thermal gradients appears to be the most effective transport mode. Calculations show that, in the case of millimetre-size droplets, Marangoni convection can significantly modify the spreading law (exponent n and constant K in $R^n \approx Kt$) established for the purely diffusive regime. However, before concluding on the relative importance of the two transport mechanisms, i.e. convection and diffusion, current models have to be improved to take into account other phenomena likely to modify the values of n and K, namely some delocalization of the reaction ahead of the triple line (for instance by evaporation/condensation) or behind the triple line (by diffusion through the reaction product layer).

The authors acknowledge the significant contributions to different parts of this study made by Professor A. Mortensen (E. P. Lausanne), Dr R. Voitovich (from the Institute of Materials Science, Kiev), Dr K. Landry and Dr C. Rado (Grenoble). We thank Dr D. Camel (CEA-Grenoble) for fruitful discussions on the topic. The contribution of J. P. Garandet and B. Drevet was performed within the frame of the GRAMME agreement between the CNES and the CEA.

Appendix A. List of symbols

a	radius of the region where $C = C_e$
C	concentration of reactive species (mass fraction)
C_e	equilibrium concentration of reactive species (mass fraction)
C_0	concentration of reactive species in the bulk (mass fraction)
C_P	mass fraction of reactive species of the reaction product
D	diffusion coefficient in the liquid
e	thickness of the reaction product layer at the triple line
F	constant close to 0.04
F_d	wetting driving force
g	gravity acceleration
G	thermal gradient
Gr	Grashof number
h	height of the drop at its centre
h'	height of the drop at position δ
J	mass flux
J_t	total mass flux
K	constant of $R^n \approx Kt$ law
L	liquid
n	exponent of $R^n \approx Kt$ law
n_v	number of moles of reactive species per unit volume of the reaction product
P	reaction product
r	radius in cylindrical coordinates
R	drop base radius
R_0	initial drop base radius
Re_M	Reynolds Marangoni number
S	solid
Sc	Schmidt number
t	time
U	convective velocity at the L–V interface
V	vapour
V_d	drop volume
V_{eff}	effective velocity
V_F	velocity of the fluid
V_{TL}	velocity of the triple line
x	one-dimensional coordinate
β_T	thermal expansion coefficient
δ	thickness of the solute boundary layer
ΔG_r	change in Gibbs energy of the reaction
Φ	angle in cylindrical coordinates
ν	kinematic viscosity of the fluid
θ	contact angle
θ_F	final contact angle
ρ_L	mass density of the liquid
ρ_P	mass density of the reaction product
σ	surface energy
τ	time scale of reactive spreading
σ'_T	derivative of σ with respect to temperature

References

Aksay, I., Hoye, C. & Pask, J. 1974 *J. Phys. Chem.* **78**, 1178.

Alboussiere, T., Neubrand, A. C., Garandet, J. P. & Moreau, R. 1997 *J. Crystal Growth* **181**, 133.

Birikh, R. V. 1966 *J. Appl. Mech. Tech. Phys.* **3**, 43.

de Gennes, P. G. 1985 *Rev. Mod. Phys.* **57**, 289.

Dezellus, O., Hodaj, F. & Eustathopoulos, N. 1998 *Proc. 2nd Int. Conf. High Temperature Capillarity, Cracow, Poland, 1997* (ed. N. Eustathopoulos & N. Sobczac). (In the press.)

Drevet, B., Landry, K., Vikner, P. & Eustathopoulos, N. 1996 *Scripta Mater.* **35**, 1265.

Eremenko, V., Kostrova, L. & Lesnic, N. 1995 *Proc. Int. Conf. High Temperature Capillarity, Smolenice Castle, Slovakia, 1994* (ed. N. Eustathopoulos), p. 113.

Espié, L., Drevet, B. & Eustathopoulos, N. 1994 *Metall. Trans.* A **25**, 599.

Eustathopoulos, N. 1996 *Proc. 'Interface Science and Materials Interconnection' (JIMIS-8)*, pp. 61–68. The Japan Institute of Metals.

Eustathopoulos, N. & Drevet, B. 1993 *MRS Symp. Proc.* **314**, 15.

Eustathopoulos, N. & Drevet, B. 1994 *J. Physique* **4**, 1865.

Fujii, H., Nakae, H. & Okada, K. 1993 *Metall. Trans.* A **24**, 1391.

Garandet, J. P. 1993 *J. Crystal Growth* **131**, 431.

Kritsalis, P., Drevet, B., Valignat, N. & Eustathopoulos, N. 1994 *Scripta Metall. Mater.* **30**, 1127.

Landry, K. & Eustathopoulos, N. 1996 *Acta Metall. Mater.* **44**, 3923.

Landry, K., Rado, C. & Eustathopoulos, N. 1996 *Metall. Mater. Trans.* A **27**, 3181.

Landry, K., Rado, C., Voitovich, R. & Eustathopoulos, N. 1997 *Acta Metall. Mater.* **45**, 3079.

Laurent, V. 1988 Thesis, Institut National Polytechnique de Grenoble, France.

Loehman, R. & Tomsia, A. 1988 *Am. Ceram. Bull.* **67**, 375.

Mortensen, A., Drevet, B. & Eustathopoulos, N. 1997 *Scripta Mater.* **36**, 645.

Mortimer, D. & Nicholas, M. 1970 *J. Mater. Sci.* **5**, 149.

Mortimer, D. & Nicholas, M. 1973 *J. Mater. Sci.* **8**, 640.

Naidich, Y. 1981 *Progress in surface and membrane science* (ed. D. A. Cadenhead & J. F. Danielli), p. 353. New York: Academic.

Nicholas, M. 1986 *Br. Ceram. Trans. J.* **85**, 144.

Nicholas, M. & Peteves, S. 1995 *Proc. Int. Conf. High Temperature Capillarity, Smolenice Castle, Slovakia, 1994* (ed. N. Eustathopoulos), p. 18.

Nogi, K. 1993 *Trans. JWRI* **22**, 183.

Rado, C. 1997 Thesis, Institut National Polytechnique de Grenoble, France.

Tanner, L. 1979 *J. Phys.* D **12**, 1473.

Voitovich, R., Mortensen, A. & Eustathopoulos, N. 1998 *Proc. 2nd Int. Conf. High Temperature Capillarity, Cracow, Poland, 1997* (ed. N. Eustathopoulos & N. Sobczac). (In the press.)

Marangoni convection in multiple bounded fluid layers and its application to materials processing

BY D. JOHNSON[1] AND R. NARAYANAN[2]

[1] Microgravity Research Centre, Université Libre de Bruxelles,
1050 Brussels, Belgium
[2] Department of Chemical Engineering, University of Florida,
Gainesville, FL 32611, USA

A brief review of multilayer convective phenomena that is associated with materials processing is presented. Several instability phenomena that can occur in a bilayer of two fluids heated from either above or below and the effect of laterally and vertically confined geometries are explained. In particular it is shown that such confinement can lead to the occurrence of codimension-two points and pure thermal coupling that is initiated by convection in an upper gas phase during liquid–gas bilayer convection. Experimental evidence that shows the effect of geometrical restrictions is given.

Keywords: Rayleigh; Marangoni; interfacial tension driven convection; buoyancy driven convection; multiple fluid level convection

1. Introduction and physics

Much of the work reported in this paper has been motivated by the need to understand a technique for growing certain crystalline materials, known as the liquid-encapsulated vertical Bridgman (LEVB) crystal growth method. Liquid-encapsulated crystal growth is a process for producing III-V semiconductor crystals from bulk liquid melts. The demand for crystals of increasingly higher purity and lower defects requires us to understand this process in much greater detail. Some examples of crystals grown using this technique are gallium arsenide (GaAs) and indium phosphide (InP). Taking GaAs as an example, when GaAs is melted, it has a tendency to decompose, releasing arsenic gas and destroying the desired stoichiometric ratio. To prevent this decomposition, a liquid encapsulant of boric oxide (B_2O_3) is placed on top of the gallium arsenide. In addition, an inert gas may be placed on top of the B_2O_3. These three layers are placed in a crucible, which is lowered through a temperature gradient created by a furnace. The lower end of the crucible is cooled, thereby solidifying the gallium arsenide. This configuration is shown schematically in figure 1. The heating configuration generates vertical as well as radial temperature gradients and, consequently, interfacial-tension-gradient-driven convection, also known as Marangoni convection, and buoyancy-driven convection, also called Rayleigh convection, can occur at the liquid–gas and liquid–liquid interfaces as well as in the bulk fluid regions.

While the LEVB technique is the motivation for this study, only by considering simple systems can we have a clearer understanding of the physics of the convective process. Radial gradients of temperature, creeping of the encapsulant along the vertical sidewalls and solutal gradients all have a complicated effect on the convection.

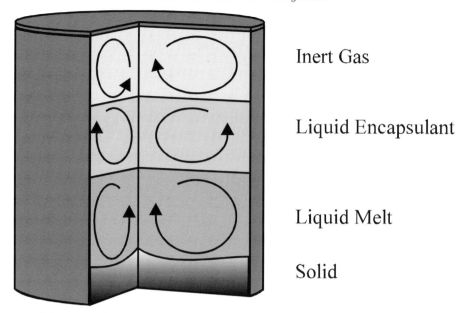

Figure 1. Schematic of a liquid encapsulated crystal grower: a system of three convecting fluid
layers. Convection in the GaAs liquid influences the quality of the GaAs solid.

Indeed the onset of convection in an actual LEVB system occurs simultaneously
with the application of any temperature gradient. However, a clear understanding
of convection in LEVB and many other materials processing methods requires us
to consider problems where classical fluid mechanical procedures may be employed,
thereby simplifying the mathematics while simultaneously revealing the essential
physical features.

One such problem is the Rayleigh–Marangoni problem. Here a gas or another liquid
superimposes a liquid layer and a vertical temperature gradient is applied. Suppose
that the density and interfacial tension of the liquid decreases with increasing temperature. As the liquid is heated from below, it is top heavy. A small disturbance
can upset this arrangement if the overall temperature difference is large enough and
flow can ensue in the form of buoyancy or Rayleigh convection. However, flow can
occur even in the absence of gravity. For example, in the quiescent state the liquid–
gas interface is flat and a small disturbance to it causes a transverse temperature
gradient at the interface causing fluid to flow from warm regions of low interfacial
tension to cold regions of high interfacial tension. Hot fluid from below rises to the
interface and cold fluid from the interface moves down to maintain continuity of fluid
flow and the convection continues as Marangoni convection. For small values of the
vertical temperature gradient, the fluids remain quiescent and transport heat by pure
conduction. However, when the temperature gradient reaches a critical value, even
the smallest disturbances imposed on the system amplify with time and the system
reaches a steady or time-periodic steady state. In other words a critical temperature
gradient is required for convection to occur. More details on the nature of this type
of convection are available in the reviews of Koschmieder (1993) and Davis (1987).
We will explain the physics of single and multilayer convection in laterally bounded
geometries where the layers are heated from above making them gravitationally stable and where the layers are heated from below making them gravitationally unstable.
The explanation of physics in multilayers will be followed by a report on two sets of

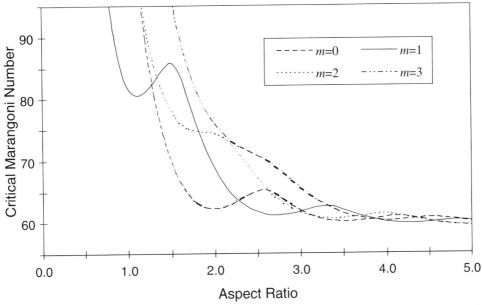

Figure 2. Plot of the critical Marangoni number versus the aspect ratio of a cylinder. The mode, m, with the smallest Marangoni number at a given aspect ratio is the mode or flow pattern at the onset of convection.

experiments. Noticeably absent from this paper will be the effects of solutal convection, some aspects of which have been covered by several other authors (McFadden *et al.* 1984; Turner 1985). It should be noted that this paper is a brief report of the work done by us and a few other researchers. Greater detail is available in the thesis by Johnson (1997) as well as papers by Johnson & Narayanan (1996, 1997, 1998). Fuller explanations of the effects of convection on crystal growth are given by Hurle (1994), Müller (1988) and Schwabe (1981).

The extent of convection is often characterized by a dimensionless temperature difference represented by the Marangoni or Rayleigh numbers. The Marangoni number is proportional to the depth of the liquid, the temperature difference and the variation of the surface tension with respect to the temperature and inversely proportional to the dynamic viscosity and thermal diffusivity. The Rayleigh number is proportional to the cube of the liquid depth, gravity, the temperature difference and the thermal expansion coefficient and inversely proportional to the kinematic viscosity and the thermal diffusivity. In a physical system, fixing the temperature difference necessarily fixes both the Rayleigh and Marangoni numbers.

We begin by confining our discussion to a single layer of fluid, heated from below, with a free surface. In this configuration both buoyancy and interfacial-tension forces become important. For larger depths, buoyancy is more important than interfacial-tension effects, and when the fluid depth is small, interfacial-tension forces dominate convection.

(a) *Physical effects of a bounded geometry*

Consider a single layer of fluid bounded below by a rigid conducting plate and whose upper surface is bounded by a passive gas. By a passive gas we mean a gas which has no viscosity and only conducts heat away. The lower plate here is at a higher temperature than the passive gas. In a fluid of infinite horizontal extent,

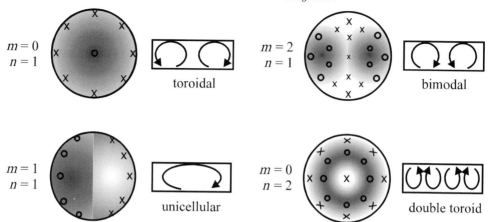

Figure 3. Schematic of four different flow patterns: ○, fluid flowing up; ×, fluid flowing down.

there is no limit on the number of convection cells. However, in a bounded finite-sized container only a finite number of convection cells may exist. Physically this means that at the onset of convection in a bounded cylinder, only one flow pattern will usually exist. As the aspect ratio (radius/height) of the container increases, more convection cells will appear. Figure 2 is a representative calculation of the critical Marangoni number for various aspect ratios and for different azimuthal modes m. The Biot number, which is a dimensionless surface heat-transfer coefficient is equal to 0.3.

In a bounded cylinder, each flow pattern is associated with an azimuthal mode, m, and radial mode, n. For example, at an aspect ratio of 1.0 in figure 2, there is an $m = 1$, $n = 1$ flow pattern (see figure 3). For an aspect ratio of 1.5, there exists an $m = 0$, $n = 1$ flow pattern. The azimuthal mode is the number of times the azimuthal component of velocity goes to zero, and the radial mode is the number of times the radial component of velocity goes to zero starting from the centre for a given vertical cross-section.

At particular aspect ratios, where the fluid switches from one flow pattern to the next, there coexist two different flow patterns. These aspect ratios are known as codimension-two points. For certain codimension-two points, the flow patterns will interact nonlinearly to yield oscillatory behaviour (Rosenblat et al. 1982; Johnson & Narayanan 1996). This phenomenon will be shown later in §2.

(b) Physical effects of multiple fluid layers

Imagine a less dense immiscible layer of fluid above the lower layer of fluid. Here the lower layer is bounded below by a rigid conducting plate and another rigid conducting plate bounds the upper layer. Once again let the temperature of the lower plate be greater than the upper plate. The interface between the two fluids may deform and is capable of transporting heat and momentum from one layer to the other. We will now consider the various types of convection that can occur in a bilayer of two fluids.

In order to distinguish the various convection mechanisms, we introduce phrases such as 'convection initiating in one layer or another'. Strictly speaking, convection occurs in both fluids simultaneously, although one layer may be more unstable than the other, driving flow in the other layer.

Turning now to various convective mechanisms, consider figure 4. Suppose that convection initiates in the lower layer. The upper layer responds by being dragged,

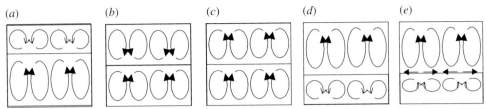

Figure 4. Schematic of the different types of convection-coupling: (*a*) lower dragging mode; (*b*) viscous coupling; (*c*) thermal coupling; (*d*) upper dragging mode; (*e*) pure thermal coupling. Moving from (*a*) to (*e*), the buoyancy force in the upper layer increases. Gas–liquid thermal coupling, with surface-driven flow, is caused by the upper fluid buoyantly convecting and simultaneously inducing interfacial-tension- or buoyancy-driven convection in the lower layer near the interface.

generating counter rolls at the interface. Hot fluid flows up in the lower layer and down in the upper layer. The upper layer is not buoyant enough and moves by a combination of viscous drag and the Marangoni effect. This is seen in figure 4a. The sign of the velocity switches and the maximum absolute value of the lower-layer velocity is much greater than the maximum absolute value of the velocity of the upper layer.

When the buoyancy in the upper layer increases and the upper layer begins to convect, one of two things can happen. The first possibility is that the two fluids are *viscously coupled.* Physically this can be shown in figure 4b as counter-rotating rolls in the two fluids. This can also be denoted by the vertical component of velocity switching sign at or near the interface, while the temperature perturbations indeed switch sign at the interface itself. If the temperature perturbation switches sign near the interface in either layer near the interface we would say that the bilayer is nearly viscously coupled. In particular if the switch takes place in the upper fluid near the interface, then the lower layer is slightly more buoyant. If the temperature perturbation switches sign in the lower layer, then the upper layer is more buoyant. The Marangoni phenomenon, for fluids, whose interfacial tension decreases with an increase in temperature, plays an ambiguous role here. The hot fluid flowing up in the lower layer causes the fluid at the interface to move in the same direction. However, the colder fluid moving down in the upper layer contradicts this. The exact effect the Marangoni phenomenon has on the two fluids depends on where the thermal perturbations change sign. For the situation where the thermal perturbations switch sign at the interface there is no Marangoni effect.

The second possibility is *thermal coupling* where the rolls are corotating (figure 4c). Here hot rising fluid from the lower layer causes hot fluid in the upper layer to flow up. The maximums of the vertical component of velocity and the temperature perturbations have the same sign in each fluid layer. Strictly speaking, the transverse components of velocity should be zero at the interface. However, thermal coupling is sometimes referred to the case when a small roll develops in one of the layers so as to satisfy the no-slip condition at the interface. In this situation, when the interfacial tension decreases with an increase in temperature, the Marangoni effect encourages flow in the lower fluid layer, and discourages the flow in the upper fluid.

Another interesting phenomenon is present at certain fluid depths where both thermal and viscous coupling can occur. At these depths, a competition arises between the two types of convection. As both convection configurations cannot occur simultaneously, the fluids begin to oscillate between these two states. This phenomenon was

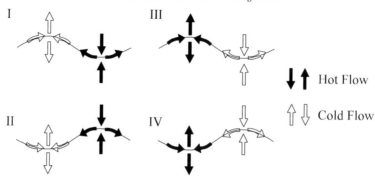

Figure 5. The four possible interfacial structures at a fluid–fluid interface. Each structure can give information about the driving force of the convection.

first reported by Gershuni & Zhukhovitskii (1982), and has recently been confirmed by Andereck *et al.* (1996).

As the buoyancy continues to increase in the upper layer, convection initiates in only the upper layer and the lower layer is viscously dragged (figure 4*d*). This situation only occurs when the upper fluid is a liquid, as gases are very tenuous and will not exert much shear. The vertical component of velocity in this case switches sign and the magnitude of convection in the upper fluid is much greater than the magnitude of convection in the lower fluid.

The last figure (figure 4*e*) is an example of what may be called *pure thermal coupling*. This typically occurs in a liquid–gas system where buoyancy convection is predominant in the gas layer. The convecting gas then simultaneously creates a non-uniform temperature profile across the liquid–gas interface and generates either Marangoni or buoyancy-driven convection in the lower layer (Johnson *et al.* 1998). Notice that the convection in the lower layer is now generated purely by horizontal temperature gradients at the interface and not by viscous dragging. To maintain the no-slip condition at the interface a small counter-roll may develop in the gas-phase. This roll is not shown in figure 4*e*.

(*c*) *Physics of interfacial structures*

Another indicator of what is occurring in bilayer convection can be inferred from the fluid–fluid interface instead of the bulk convection. In a paper by Zhao *et al.* (1995), four different interfacial structures were identified for any given convecting bilayer with a deflecting interface. Each of these structures depends upon whether fluid was flowing into or away from the trough or the crest, and whether the fluid was hotter or cooler at the trough or the crest of the interface. Hot fluid flowing into a trough defines the first interfacial structure. The second interfacial structure has hot fluid flowing into a crest. The third structure has hot fluid flowing away from a crest and the fourth structure has hot fluid flowing away from a trough. Each of these four scenarios is given in figure 5.

One of the important factors to consider in interfacial structures is the direction of the flow along the interface. As interfacial tension is usually inversely proportional to temperature, at cooler regions of the interface, the interfacial tension will be higher and will pull on the interface. Where the interface is hotter, the interfacial tension will be lower causing the fluid to move away from warmer regions. Another important factor is the direction of the flow into or away from a crest or a trough. One reason the interface deflects is due to bulk convection, caused by buoyancy effects, pushing against the interface. Consider two fluids whose dynamic viscosities are equal. If buoyancy-driven convection is occurring mostly in the lower layer, then the fluid will flow up from the lower layer into a crest. If the fluid flows down from the top layer into a trough, then one would argue that buoyancy-driven convection occurs mostly in the upper fluid.

In each of the four cases, the interfacial structure can be used to indicate the driving force of the convection. In the first interfacial structure, the dominating driving force is interfacial-tension-gradient-driven convection. This is seen as the cold fluid, with the higher interfacial tension pulling the fluid up into the crest. The first interfacial structure can also occur by buoyancy-driven convection in the upper layer, when the density of the upper layer increases with an increase in temperature. In the second interfacial structure, buoyancy drives convection in the lower phase. The hot rising fluid pushes the interface upwards. As the fluid moves along the interface, it cools and eventually sinks back down. The third interfacial structure is dominated by buoyancy-driven convection in the upper phase or by interfacial-tension-driven convection where the interfacial tension increases with respect to temperature. The fourth interfacial structure only occurs when the lower fluid has a positive thermal-expansion coefficient. In other words, the density increases with an increase in the temperature, causing the cooler lower fluid to flow up into a crest.

Knowledge of interfacial structures will be beneficial in the understanding of certain materials processing problems such as drying of films, coatings and deposition.

(d) Physics of heating from above

In the previous subsections we talked about some of the phenomena that occur in single- and multiple-fluid layers heated from below. However, in an attempt to avoid convection in crystal growth, the crucible is often cooled from below and heated from above. This heating configuration changes the physics of the problem, which is the topic of this subsection.

When a layer of fluid is being heated from above, it creates a stable density stratification. Therefore not only does the buoyancy force not cause convection, it acts to inhibit other instabilities. Marangoni convection, though, may still occur in fluids being heated from above.

First we will consider a single layer of fluid superposed by a passive gas. If the upper gas is truly passive, then pure Marangoni convection will not occur. For example, suppose some random perturbation causes some part of the surface to become warmer than the rest of the surface. The interfacial tension will decrease in this region and the tension will pull fluid away from this hot spot. By continuity, fluid lying below the hot spot will rise up to replace the displaced fluid. As the lower fluid is cooler than the surface this region now cools off and the interfacial tension increases, thereby restabilizing the region. However, in real systems, the upper fluid is never truly passive. Given the same scenario, fluid movement along the interface will also drag warmer fluid from above. This warmer upper fluid will further increase the temperature in this region, and, depending upon the ratio of thermal–physical properties, reinforces the instability.

By this argument, it appears that an active upper fluid is necessary to have Marangoni convection when the system is being heated from above. However, this is not the case if the buoyancy effects are included. In Rednikov *et al.* (1998), it was demonstrated, theoretically, that oscillatory onset of convection may occur for a single layer of fluid with a purely passive upper gas. The explanation is as follows.

If a small volume of fluid is displaced within the bulk of the fluid, the density stratification acts as a restoring force, causing a dampened oscillation within the fluid. These are often referred to as internal waves. The Marangoni force acts similarly, as discussed above, also giving dampened oscillations. Apparently when these two forces act together they can overshoot one another leading to sustained oscillatory convection. Indeed, as was demonstrated in their paper, this only occurs in certain fluids, at certain depths, where the buoyancy and interfacial-tension forces are approximately equal.

A completely different type of instability is also possible in two layers of fluids being heated from above. Gershuni & Zhukhovitskii (1981) first demonstrated this phenomenon by analysing two immiscible fluids where the interface between the fluids was assumed flat and the Marangoni phenomenon was neglected. They found the onset of steady convection when the thermal conductivity and thermal expansivity of the lower fluid was much greater than that of the upper fluid.

The mechanism of this instability is as follows. Suppose an element of fluid in the upper layer, near the interface, is displaced towards the lower layer. Because the thermal expansion of the upper fluid is so small, this element remains in a relatively neutrally buoyant state. Also, as the thermal conductivity of the upper fluid is small, it cools very slowly. When the two fluids are heated from above, the displaced fluid will be warmer than its surroundings. This element of fluid then heats part of the lower fluid near the interface. The lower fluid, with its relatively large thermal conductivity and thermal expansivity, quickly heats up and expands horizontally. This expansion then causes convection in the lower fluid layer and propagates by viscously coupling with the upper fluid layer. The Marangoni phenomenon, if it were considered, would act to enhance this instability.

Another case of interest is convection induced by the Rayleigh–Taylor instability. This phenomenon can occur in two immiscible fluid layers being heated either from above or below, when the densities of the two fluids are approximately the same and the thermal expansivity of the lower fluid is much greater than the thermal expansivity of the upper fluid. Upon heating, the density of the lower fluid will decrease and become less than the density of the upper fluid. Consequently, the heavier upper fluid will begin to sink causing large deformations in the liquid–liquid interface, generating the Rayleigh–Taylor instability (Chandrasekhar 1961). This problem has been investigated extensively in Renardy & Renardy (1985) and Renardy (1996), but is of application to materials processing only if the densities of both layers are similar.

2. Some experimental observations

The experiments were used to investigate both the oscillatory behaviour near codimension-two points and the pure thermal coupling of air with silicone oil (see figure 4e). Details on the experimental procedure are available in Johnson (1997) and Johnson & Narayanan (1996, 1998).

(a) Experimental apparatus and procedure

The experiments consisted of two compartments: one for the lower fluid and one for the air. Lucite inserts were used to give a variety of different fluid depths and aspect ratios. A copper plate was placed below the liquid insert and the air insert was bounded above by a high-thermal-conductivity infrared transparent zinc selenide (ZnSe) window. Heating of the copper plate was done by an enclosed stirred water bath that was in turn heated by a hot plate. The top of the ZnSe window was kept at a constant temperature by accurately controlling the temperature of the overlying air. The overall temperature control was kept within ±0.05 °C.

The flow patterns that developed at the silicone oil–air interface layer were visualized with an infrared camera. Although other flow visualization techniques could have been used, such as shadowgraphy or particle tracing, the IR camera was chosen to prove the viability of its use with opaque materials, such as gallium and gallium arsenide. The IR imaging technique is also useful in observing weak thermocapillary flow near the surface, whereas shadowgraphy requires some strength in the domain flow.

To guarantee that the flow pattern seen was indeed the flow pattern at the onset of convection, the temperature difference applied across the bilayer system was carefully increased. At first a temperature difference was applied that was less than the critical temperature difference necessary for the onset of convection. This, and all temperature differences, were kept constant for several characteristic time constants; *ca.* 3–4 h. If no flow pattern was seen, the temperature difference was then increased by as little as 0.05 °C. This was repeated until the temperature profile at the interface changed to some distinct pattern, indicating that the fluid had begun to flow. At this point, the flow pattern was recorded and the temperature difference noted.

(b) *Experimental observation of codimension-two points*

As was noted in §1, there exist certain liquid aspect ratios where two different flow patterns coexist. For example, in figure 2 at an aspect ratio of 2.3, there exists a codimension-two point between the azimuthal modes $m = 0$ and $m = 1$. The questions we want to answer are: What happens at these aspect ratios? Does one flow dominate over the other? Do the different flow patterns coexist as a superposition of both states, or do they oscillate and interact between these two states?

Rosenblat *et al.* (1982) have performed a weakly nonlinear analysis to investigate these questions. They found that all three of these possibilities may occur, depending on the Prandtl number, the particular codimension-two point being investigated, and on which side of the codimension-two point the aspect ratio lies. To simplify their calculations, a vertical and tangential vorticity-free side-wall was assumed. Later a more realistic no-slip condition was applied (Zaman & Narayanan 1996; Dauby *et al.* 1997), where it was noted that the order of azimuthal modes, as the aspect ratio was increased, was different than the vorticity-free condition. These latter calculations were done assuming a linearized instability analysis. Therefore, a direct comparison of the nonlinear analysis with the experiment is not currently possible. Nonetheless, some of the qualitative features should still hold true.

A series of experiments were performed to first find the codimension-two points and then determine the flow patterns at or near the codimension-two point (Johnson & Narayanan 1996). A 5.0 mm-high 2.5 aspect-ratio liquid insert was used in conjunction with a 11.2 mm air height. Table 1 shows the calculated critical Marangoni numbers for four different azimuthal modes for a 2.5 aspect ratio. The table predicts that an $m = 0$ flow pattern should be seen at the onset of convection. However, the critical Marangoni numbers for $m = 1$ and $m = 2$ are also very close to the onset point. Physically, this means that for temperature differences slightly above the critical temperature difference, these modes may affect the flow pattern.

At the onset of convection, a very faint $m = 0$ double-toroidal-flow pattern could be seen. When the temperature difference was increased by just 0.05 °C, the flow

Table 1. *Critical Marangoni number for the first four azimuthal modes for a 2.5 aspect ratio.*

modes	Marangoni number
0	69.37
1	70.84
2	70.41
3	72.98

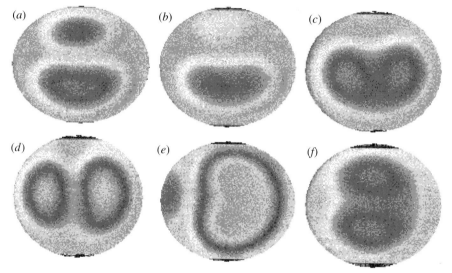

Figure 6. Infrared images showing the mode-switching behaviour in the paper by Johnson & Narayanan (1996). The experiment used 91 cS silicone oil and a 5.0 mm, 2.5 aspect-ratio insert.

pattern changed from the double toroid to a dynamic switching between two and one flow cells (see figure 6).

At first, two symmetric cells appeared (figure 6a). Then, one of the cells would grow and push the other cell out of the picture, forming a superposition of the $m = 0$ and $m = 1$ flow pattern (figure 6b). Next one cell would grow (figure 6c), then split into two cells, rotated by 90° (figure 6d). This process would then repeat itself (figure 6e) arriving back to the original $m = 2$ flow pattern (figure 6f). As long as the temperature difference was held constant, this dynamic process would continue repeating itself approximately every 20 min.

It is noteworthy that oscillatory convection is of particular importance in crystal growth. It has been shown (Hurle 1994) that fluctuating temperatures in the liquid melt have a deleterious effect on the crystal quality, leading to a higher dislocation density.

(c) Experimental observations of thermal coupling

The thermal coupling of air with the lower fluid was originally discovered by a series of experiments using the same experimental apparatus (Johnson & Narayanan 1998). As was explained in § 1, air can thermally couple with the lower fluid caus-

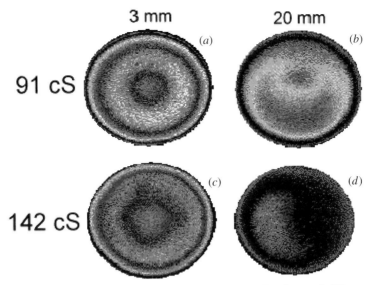

Figure 7. Infrared images of the flow pattern for different air heights and different viscosities of silicone oil. (*a*) and (*b*) used a 91 cS silicone oil. (*c*) and (*d*) used a 142 cS silicone oil. (*a*) and (*c*) had a 3 mm air height, (*b*) and (*d*) had a 20 mm air height.

ing interfacial-tension-driven flow in the lower fluid. To explore this, a set of four experiments was performed.

In all of the experiments, a liquid aspect ratio of 2.0 was studied. Two different air heights (3 and 20 mm) and two different viscosities were used. When the air height of 3 mm was used, the flow patterns did not change with viscosity but the temperature difference across the liquid did increase proportionally, indicating that convection was controlled by the liquid phase. The liquid convection pattern also agreed with calculations and the experimental result is seen pictorially in figures 7*a*, *c*.

When a deeper air height was used, the temperature difference across the liquid at the onset of convection did not change substantially between the experiments that employed different fluid viscosities. This indicated that convection was controlled by the dynamics in the air layer. It may be argued that air convection, when dominant, acts like buoyancy-driven convection between two rigid conducting plates as the lower liquid is much more viscous and more conductive than the air above it. A comparison was therefore made between the measured temperature drops across the air for the deeper air heights and the numerical calculations of Hardin *et al.* (1990). The experimental and theoretical results compared remarkably well and the flow pattern predicted theoretically also compared favourably with the experimental results. This confirmed our hypothesis that deep air heights interact with the lower liquid and drive thermally coupled flow through the interface.

This phenomenon of thermal coupling may not be as important in LEVB because the crucible is often heated from above. However, this may be much more applicable to other important processes, such as evaporation and drying of films.

3. Future work

The research of convection in multiple fluid layers has revealed and continues to reveal many interesting phenomena. However, further work is needed to elucidate some of the details more fully in a realistic system. One of the more interesting areas involves analysing some of the basic instability phenomena in bounded containers. To do this, two-fluid-layer numerical models that take into account realistic no-slip conditions will be necessary. With a proper model some of the effects, such as the Rayleigh–Taylor instability, the Gershuni–Zhukhovitskii instability and oscillations between thermal and viscous coupling, can be studied for containers with small aspect ratios.

To date, few experiments have been performed in small aspect-ratio containers. As was demonstrated in the codimension-two point experiments, new and interesting dynamics are present in small aspect-ratio containers, which are not present in large aspect-ratio containers. It would be interesting to show the interaction of codimension-two points with such instabilities as the oscillations between thermal and viscous coupling. Additionally, several of the phenomena discovered with theoretical models have yet to be shown in experiments. Two examples are the Gershuni–Zhukhovitskii instability and the oscillations shown by Rednikov *et al.* (1998).

By investigating some of the basic physics of multilayer convection, we obtain a better understanding and appreciation for the liquid-encapsulated crystal-growth process and other fluid materials processing problems where temperature gradients are employed. Further research into this field should lead to improvements in such an important industrial process.

We are grateful to NSF for their support via grants CTS 9307819 and CTS 95-00393 and to NASA via grants NGT 3-52320 and NAG 1-1474. The authors thank Pierre Dauby for figure 2.

References

Andereck, C. D., Colovas, P. W. & Degen, M. M. 1996 Multilayer convection. *Proc. of the AMS-IMS-SIAM Joint Research Conf. on Multifluid Flows*, Philadelphia, PA: SIAM.

Chandrasekhar, S. 1961 *Hydrodynamic and hydromagnetic stability*. Oxford University Press

Dauby, P. C., Lebon, G. & Bouhy, E. 1997 Linear Béenard–Marangoni instability in rigid circular containers. *Phys. Rev.* E **56**, 520.

Davis, S. H. 1987 Thermocapillary instabilities. *A. Rev. Fluid Mech.* **19**, 403.

Gershuni, G. Z. & Zhukhovitskii, E. M. 1981 Instability of a system of horizontal layers of immiscible fluids heated from above. *Izv. Akad. Nauk. SSSR, Mekh. Gaza* **6**, 28.

Gershuni, G. Z. & Zhukhovitskii, E. M. 1982 Monotonic and oscillatory instabilities of a two-layer system of immiscible liquids heated from below. *Sov. Phys. Dokl.* **27**, 531.

Hardin, G. R., Sani, R. L., Henry, D. & Roux, B. 1990 Buoyancy-driven instability in a vertical cylinder: binary fluids with Soret effect. I. General theory and stationary stability results. *Int. J. Numer. Meth. Fluids* **10**, 79.

Hurle, D. T. J. (ed.) 1994 *Handbook of crystal growth*. Amsterdam: North-Holland.

Johnson, D. 1997 Geometric effects on convection in cylindrical containers. Ph.D. thesis, University of Florida.

Johnson, D. & Narayanan, R. 1996 Experimental observation of dynamic mode switching in interfacial-tension-driven convection near a codimension-two point. *Phys. Rev.* E **54**, R3102.

Johnson, D. & Narayanan, R. 1997 Geometric effects on convective coupling and interfacial structures in bilayer convection. *Phys. Rev.* E **56**, 5462.

Johnson, D., Narayanan, R. & Dauby, P. C. 1998 The effect of air on the pattern formation in liquid–air bilayer convection: how passive is air? *Phys. Fluids* (Submitted.)

Koschmieder, E. L. 1993 *Bénard cells and Taylor vortices*. Cambridge University Press

McFadden, G. B., Coriell, S. R., Boisvert, R. F., Glicksman, M. E. & Fang, Q. T. 1984 Morphological stability in the presence of fluid flow in the melt. *Metal. Trans.* A **15**, 2117.

Müller, G. 1988 *Crystal growth from the melt*. Berlin: Springer.

Rednikov, A., Colinet, P., Velarde, M. G. & Legros, J. C. 1998 Rayleigh–Marangoni oscillatory instability in a horizontal liquid layer heated from above: coupling between internal and surface waves. *J. Fluid Mech.* (Submitted.)

Renardy, Y. Y. 1996 Pattern formation for oscillatory bulk-mode competition in a two-layer Bénard problem. *Z. Angew Math. Phys.* **47**, 567.

Renardy, Y. Y. & Renardy, M. 1985 Perturbation analysis of steady and oscillatory onset in a Bénard problem with two similar liquids. *Phys. Fluids* **28**, 2699.

Rosenblat, S., Davis, S. H. & Homsy, G. M. 1982 Nonlinear Marangoni convection in bounded layers. I. Circular cylindrical containers. *J. Fluid Mech.* **120**, 91.

Schwabe, D. 1981 Marangoni effects in crystal growth melts. *Physicochem. Hydrodyn.* **2**, 263.

Turner, J. S. 1985 Multicomponent convection. *A. Rev. Fluid Mech.* **17**, 11.

Zaman, A. & Narayanan, R. 1996 Interfacial and buoyancy-driven convection: the effect of geometry and comparison with experiments, *J. Colloid Interf. Sci.* **179**, 151.

Zhao, A. X., Wagner, C., Narayanan, R. & Friedrich, R. 1995 Bilayer Rayleigh–Marangoni convection: transitions in flow structures at the interface. *Proc. R. Soc. Lond.* A **451**, 487.

Discussion

J. R. HELLIWELL (*Department of Chemistry, University of Manchester, UK*). Reference has been made to oscillatory convection flow patterns in fluids and that this is known to cause defects (dislocations and fault lines) in crystal growth. My own particular research interests include the growth of protein crystals for X-ray crystal-structure analysis and how the quality of protein crystals can be improved, and thereby exploited, for higher resolution X-ray crystallographic data collection. I have been using CCD and interferometry diagnostic monitoring of protein crystal growth, and have seen benefits of microgravity if the crystals do not move, and if the mother liquor is not subject to convection (including Marangoni convection). The benefits of these conditions manifest as reduced crystal mosaicity and likewise a lack of, or only a few, mosaic blocks in X-ray topographs of crystals in such cases. In his experiment, how can Dr Johnson be sure that it is specifically oscillatory flow patterns that especially caused defects in his type of crystal?

D. JOHNSON. We cannot be sure that oscillatory convection is always responsible for defects in crystals. However, research cited by Hurle (1994) has indicated that oscillatory behaviour generated through double diffusion is the cause of striations along the growth axis in directional solidification. The point of this paper, however, is to show that oscillatory behaviour need not arise merely from opposing forces that are seen in thermo-solutal, otherwise known as double-diffusive, convection. Such oscillatory behaviour can arise by geometrical effects. Indeed as the crystal grows, the aspect ratio of the liquid phase changes and there are certain aspect ratios where the energy states may coexist leading to codimension-two points that can cause oscillatory convection.

S. K. WILSON (*Department of Mathematics, University of Strathclyde, Glasgow, UK*). I complement the authors on a penetrating investigation of a complicated physical situation. As I understand it, they have found examples of slow oscillations between two different steady flow patterns in the vicinity of codimension-two points calculated theoretically using linear theory for the onset of steady convection in a finite-sized container. May I ask if truly oscillatory (rather than quasi-steady) convection is ever observed, and if it would be possible to undertake the same kind of investigation for the onset of oscillatory convection?

D. JOHNSON. Yes, under certain circumstances, we believe that oscillatory convection can be observed in liquid–gas bilayer experiments. Theoretical calculations that were done indicate the absence of such convection at the onset. In that case what is the origin of oscillations in our experiments? The answer lies in the fact that a Hopf bifurcation lies in the vicinity of the onset but only in the post-onset region. The Hopf bifurcation mode or oscillatory mode was excited by the presence of codimension-two points. These points were generated by the fact that at certain aspect ratios two competing flow states coexist and in a manner of speaking the system wants to choose between the flow states leading to continual oscillations.

Interfacial phenomena of molten silicon: Marangoni flow and surface tension

By Taketoshi Hibiya[1], Shin Nakamura[1], Kusuhiro Mukai[2],
Zheng-Gang Niu[2], Nobuyuki Imaishi[3], Shin-ichi Nishizawa[4],
Shin-ichi Yoda[5] and Masato Koyama[6]

[1] Fundamental Research Laboratories, NEC Corporation,
34 Miyukigaoka, Tsukuba 305-8501, Japan
[2] Department of Materials Science and Engineering, Kyushu Institute of
Technology, 1-1, Sensui-cho, Tobata-ku, Kita-Kyushu 804-8550, Japan
[3] Institute of Advanced Material Study, Kyushu University, 6-1,
Kasuga Koen, Kasuga 816-8580, Japan
[4] Frontier Technology Division, Electrotechnical Laboratory, AIST,
MITI, 1-4 Umezono 1-chome, Tsukuba 305-8568, Japan
[5] Space Experiment Department, National Space Development Agency of Japan,
1-1 Sengen 2-chome, Tsukuba 305-8505, Japan
[6] Frontier Joint Research Program Department, Japan Space Utilization Promotion
Center, 30-16, Nishi-Waseda 3-chome, Shinjuku-ku, Tokyo 169-8624, Japan

Temperature oscillation due to the oscillatory Marangoni flow was measured for a
molten half-zone silicon column (10 mm high and 10 mm in diameter with a temperature difference of 150 K between the upper and lower solid–liquid interfaces) under
microgravity by using fine thermocouples. The flow is in a hypercritical condition;
that is, the Marangoni number is estimated to be over 10 000. The structure of the
Marangoni instability is two-fold symmetry for the small aspect ratio (height/radius)
Γ of 1 and one-fold symmetry for the melt with Γ of 2. The surface tension of molten
silicon was measured by a sessile drop method in carefully controlled ambient atmospheres with various oxygen partial pressures from 4×10^{-22} to 6×10^{-19} MPa. These
measurements showed that the surface tension and its temperature coefficient showed
a marked dependence on oxygen partial pressure. Accordingly the effect of oxygen
partial pressure on the Marangoni flow should be made clear. Moreover, Marangoni
flow at the flat surface, which corresponds to the flow for the Czochralski growth
system, should also be studied.

Keywords: molten silicon; microgravity; liquid column; instability; surface tension;
oxygen partial pressure

1. Introduction

The crystal growth process of semiconductor silicon is controlled by a heat and mass
transfer process. Within the bulk melt on Earth, buoyancy flow plays an important
role, whereas at the melt surface, Marangoni flow is thought to exist and be involved
in a heat and mass transfer process. For buoyancy driven flow, the flow mode has
been experimentally characterized using an X-ray flow visualization technique and
numerical modelling (Kakimoto et al. 1988). However, existence of the Marangoni

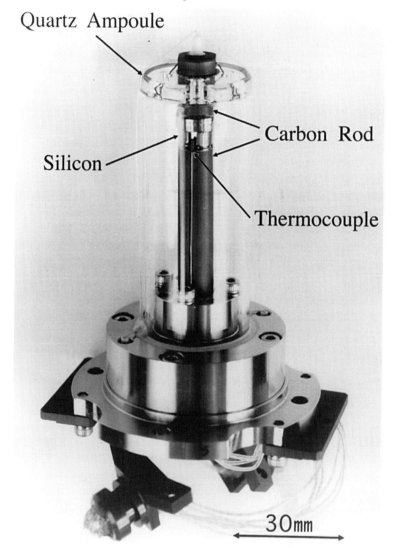

Figure 1. Ampoule containing a silicon specimen.

flow was confirmed for the first time through silicon crystal growth experiment in microgravity (Eyer *et al.* 1985). Although there have been several works reported for the Marangoni flow of other fluids with middle or high Prandtl number, e.g. silicone oil or molten salt (Schwabe *et al.* 1978; Wanschura *et al.* 1995), only limited numbers of works have been reported on fluids with low Prandtlt number, Pr, fluids including molten silicon (Cröll *et al.* 1989; Levenstam & Amberg 1995; Han *et al.* 1996).

There are two types of Marangoni flow; one occurs in a liquid column and corresponds to floating zone crystal growth, and the other occurs at a flat surface and corresponds to Czochralski crystal growth using a crucible. Research on the Marangoni flow of molten silicon has been carried out only for the liquid column configuration.

Temperature coefficient of molten silicon is a driving force for the Marangoni flow. However, surface tension and its temperature coefficient for molten silicon were suggested to be sensitive to the amount of adsorbed oxygen on the melt surface

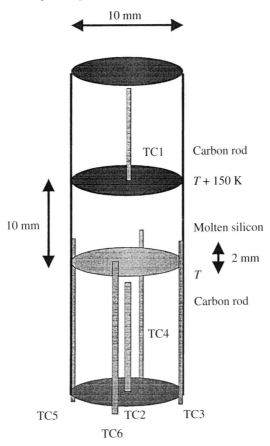

Figure 2. A sketch of the silicon specimen.

(Keene 1987). The oxygen concentration in silicon crystals is higher in the melt contained by a silica crucible, i.e. in the Czochralski case, and lower in the melt for the floating zone configuration. However, precise measurements of surface tension and its temperature dependence for molten silicon in ambient atmosphere with defined oxygen partial pressure has not yet been reported.

In the present work, we have tried to examine experimentally the Marangoni flow of molten silicon in microgravity and also on Earth. There are several approaches to Marangoni flow research, i.e. measurement of a temperature field, observation of a flow field, surface oscillation observation, crystal growth, computational simulation and so on. Among these we have employed temperature oscillation measurements and computational simulations in the present work. The effect of the oxygen partial pressure of ambient atmosphere on the temperature coefficient of the surface tension of molten silicon and on the Marangoni flow is also discussed.

2. Experiment

Figures 1 and 2 show the configuration of the present experiment. A silicon specimen 10 mm high and 10 mm in diameter was sustained between the upper and lower carbon rods within a silica glass ampoule. A molten silicon column was formed under microgravity conditions on board the NASDA (National Space Development Agency

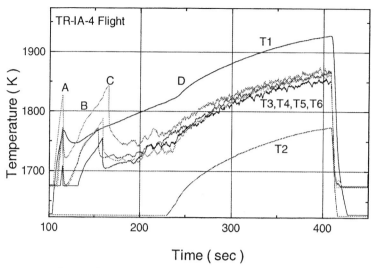

Figure 3. Temperature fluctuation in microgravity. Lamp power was changed from 1300 to 900 W at point A, where temperature detected by one thermocouples out of six reached 1823 K. Melting started at point B. Around point C temperature $(T3-T6)$ decreased because the thermocouples touched the silicon melt. At point D a liquid column was formed completely.

of Japan) sounding rocket TR-I-A #4 using a mirror furnace (Hibiya *et al.* 1996), where only Marangoni flow can be observed by eliminating the effect of buoyancy. A temperature difference of 150 K was maintained between the upper and lower rods. As shown in figure 2, four fine thermocouples separated by 90° along the azimuthal direction were attached to the melt surface in order to detect temperature oscillation due to the Marangoni instability (the sampling frequency was 40 Hz). Two thermocouples were set within carbon rods close to the solid–liquid interface to monitor the temperature difference between the upper and lower interfaces. Argon gas of 6N-purity was flowed through an ampoule at a rate of 2ℓ min^{-1} to prohibit condensation of silicon oxide on the ampoule inner wall. The detailed experimental procedure is reported elsewhere (Nakamura *et al.* 1998).

3. Results and discussion

(a) Structure of the Maranogni instability

Figure 3 shows temperature data obtained in microgravity and temperature oscillations due to instability of the Marangoni flow of the silicon melt column. As clearly seen in figures 4a, b, temperature oscillation during column formation (C-D in figure 3) shows a marked difference from that after column formation (after D). Spectral analysis showed that observed frequencies were 0.1 Hz for column formation and 0.15 Hz after column formation, while power spectrum density after column formation was too low for finding a particular frequency, as shown in figure 5. This means that flow is in a hypercritical condition, i.e. far away from a transition from a stationary to an oscillatory flow.

Comparing figures 4a, b, the phase relationship among temperature signals detected by the four thermocouples (see figure 2) is different during and after column formation. During the column formation period, temperature measured by the thermocouples opposite each other (T4F1 detected by TC4 and T6F1 detected by TC6)

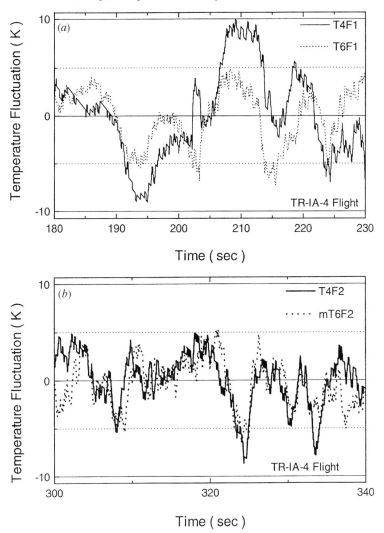

Figure 4. (*a*) Temperature fluctuation data during the column formation. T4F1 and T6F1 were obtained from the temperature data (see figure 3) by subtracting the average temperature increase. T4F1 and T6F1 correspond to the thermocouples TC4 and TC6, respectively. (*b*) Temperature fluctuation data after the column formation. The data were obtained by the same method as in (*a*). mT6F2 represents the inverse sign of T6F2.

fluctuates in the same phase as shown in figure 4*a*, while temperature fluctuation is the antiphase relation between thermocouples separated by 90°. After the column formation, temperature fluctuation data measured by the thermocouples set opposite each other show an antiphase relationship, as shown in figure 4*b*. Note that in figure 4*b* temperature fluctuation data mT6F2 represents the inverse of T6F2.

The above mentioned phase relationship of the temperature fluctuation can suggest the existence and motion of a non-axisymmetric temperature field within a liquid column. A cold temperature region is formed in the liquid column, because the liquid column is heated by radiation from the lamp. When the antiphase relationship is detected by the thermocouples set opposite each other, the existence of a one-fold symmetrical temperature field is plausible. The cold temperature region is

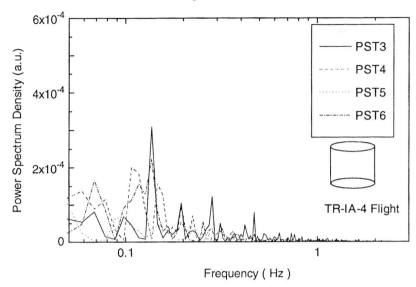

Figure 5. Power spectrum density obtained after column formation. PST3 was calculated using the temperature fluctuation data detected by thermocouple TC3.

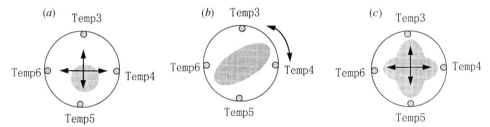

Figure 6. Proposed model of non-axisymmetric distribution of temperature in a molten silicon column: (*a*) one-fold symmetrical distribution: (*b*) and (*c*) two-fold symmetrical distribution. Temperature in a shadowed region is lower than in a surrounding region.

expected to move randomly, as shown in figure 6*a*. The antiphase relationship of the temperature fluctuation between the neighbouring thermocouples (separated by 90°) suggests the existence of a two-fold symmetrical temperature field. The structures of these temperature fields are shown in figures 6*b, c*: rotation and/or pulsation.

Wanshura *et al.* suggest that the structure of these instabilities depends on the aspect ratio of the liquid column Γ. The wave number of the Maranogni instability for liquid column m is written as $m = 2/\Gamma$ (Wanschura *et al.* 1995). This idea can explain the present experimental results: that is, $m = 2$ for the column formation period (short column) and $m = 1$ after column formation (long column). The measured frequency corresponds to that of azimuthal oscillation of the instability structure.

Numerical simulation for occurrence of the instability in the molten silicon in the present configuration is not finished, while the simulation for the liquid with a Prandtl number Pr of 1.0 in a similar configuration to the present one suggests that instabilities with $m = 1, 2$ can take place depending on the column aspect ratio. Figure 7 shows velocity and temperature fields of a liquid column with a Prandtl number of 1.02, aspect ratio of 1.33 and Marangoni number Ma of 3330 (Imaishi & Yasuhiro 1997). A cross section is shown at a melt height $Z = 0.855\Gamma$. As shown in figure 7, instability with a wave number $m = 2$ takes place for a short column. The

(a)   (b)

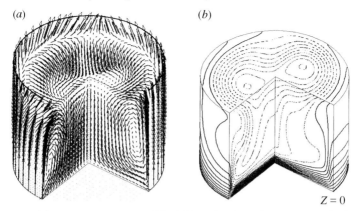

$Z = 0$

Figure 7. Velocity (*a*) and temperature (*b*) fields of a liquid column with Prandtl number Pr = 1.02, aspect ratio Γ = 1.33 and Marangoni number Ma = 3330, which shows instability with wave number of m = 2. Cross section is shown at height Z = 0.855.

simulation results support the present experimental result of temperature fluctuation measurements.

(*b*) *Oxygen partial pressure dependence of surface tension and its temperature coefficient*

As mentioned in §3*a* the Marangoni flow of the molten silicon column shows instability and the power spectrum density of the oscillation is very weak. This means that the flow is in a hypercritical condition. This is supported by a calculated Marangoni number, which is much higher than the critical Marangoni number of about 100 (Cröll *et al.* 1989), as follows:

$$Ma = (|\partial\gamma/\partial T|\Delta T L)/(\mu\kappa). \qquad (3.1)$$

Here, $\partial\gamma/\partial T$ is the temperature coefficient of surface tension for molten silicon, ΔT is the temperature difference between the upper and lower interfaces of the column, L is the column height, μ and κ are the viscosity ($\mu = 7 \times 10^{-4}$ kg m^{-1} s^{-1}) and thermal diffusivity ($\kappa = 2.12 \times 10^{-5}$ m^2 s^{-1}) of the molten silicon, respectively. The Marangoni number ranges from 5000 to 20 000 depending on $\partial\gamma/\partial T$. According to Keene, surface tension and its temperature coefficient of molten silicon depend on the magnitude of surface contamination by adsorbed oxygen (figure 8) (Keene 1987). We measured surface tension and its temperature coefficient using a sessile drop method in ambient atmospheres with various oxygen partial pressures (Niu *et al.* 1998). A substrate made of BN was used for the following reasons: this is less reactive with molten silicon and a large contact angle can be obtained so that good measurement accuracy for surface tension is assured. An electromagnetic levitation technique was also employed for surface tension measurement (Przyborowski *et al.* 1995).

Figure 9 shows surface tension of molten silicon as a function of temperature for various oxygen partial pressures of the ambient atmosphere from 4.8×10^{-22} to 1×10^{-19} MPa. It is noteworthy that surface tension and its temperature coefficient show marked dependence on oxygen partial pressure. From this figure it is understood that surface tension and its temperature coefficient measured by a levitation technique in a 6N-argon gas atmosphere ($\gamma = 783.5 \times 10^{-3}$ N m^{-1} and $\partial\gamma/\partial T = -0.65 \times 10^{-3}$ N m^{-1} K^{-1}) correspond to those measured in an oxygen partial pressure of *ca.* 10^{-21} MPa. If we plot surface tension as a function of logarithmic ambient oxygen

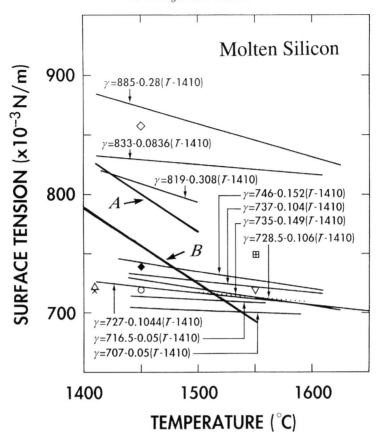

Figure 8. Reported surface tension of molten silicon reviewed by Keene (1987). Newly obtained data by the authors are also included: (A) by the sessile drop method at $P_{O_2} = 4.8 \times 10^{-22}$ MPa (Niu $et\ al.$ 1998) and (B) by the levitation method in 6N argon atmosphere (Przyborowski $et\ al.$ 1995).

partial pressure (see figure 10), the gradient corresponds to the excess amount of oxygen atoms adsorbed on the molten silicon surface (Γ_O) according to the Gibbs adsorption isotherm

$$\Gamma_O = -2(1/RT)(\partial\gamma/\partial\ln P_{O_2}). \tag{3.2}$$

Here, R and T are the gas constant and temperature, respectively. The excess amount of oxygen on the silicon melt surface at full coverage, Γ_O, was evaluated to be 2.1×10^{-6} mole m^{-2} at 1693 K and at P_{O_2} of 1×10^{-21}–5×10^{-20} MPa. This means that one out of every ten atoms is oxygen and suggests that the oxygen concentration is extremely high at the surface, while the bulk concentration of oxygen was calculated to be $ca.$ 0.001 mass% (8.5×10^{17} atoms cm$^{-3} \approx 18$ ppm), based on the equilibrium constant for the following chemical reaction between an oxygen molecule in a gas phase and an oxygen atom in molten silicon (Hirata & Hoshikawa 1990):

$$\tfrac{1}{2}O_2(g) = \underline{O}(l). \tag{3.3}$$

The temperature coefficient of molten silicon is plotted as a function of oxygen partial pressure of the ambient atmosphere (figure 11). A larger absolute value of temperature coefficient was obtained than that previously reported (compare with the data in figure 8).

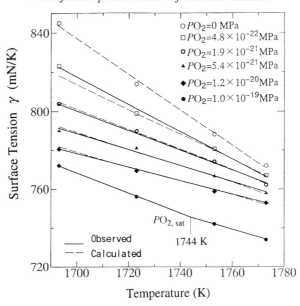

Figure 9. Surface tension of molten silicon as a function of temperature for various oxygen partial pressures of ambient atmosphere from 4.8×10^{-22} to 1×10^{-19} MPa.

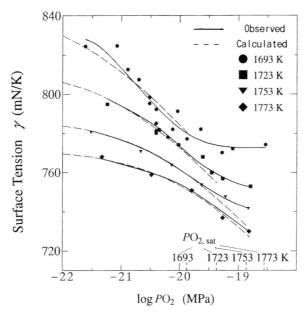

Figure 10. Surface tension of molten silicon as a function of logarithmic oxygen partial pressure of ambient atmosphere P_{O_2}.

As mentioned above the role of oxygen on surface tension and its temperature coefficient have been thoroughly clarified. However, unanswered questions still remain. Although the melt oxygen concentration for specimens which were employed for surface tension measurement by the sessile drop method and showed temperature coefficients of $-0.4 \sim -0.8 \times 10^{-3}$ N m^{-1} K^{-1} was estimated to be in the order of 0.001 mass% (5×10^{17} atoms cm^{-3} $\approx$ 10 ppm) based on thermodynamical calcula-

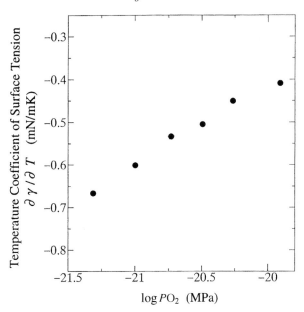

Figure 11. Temperature coefficient of surface tension of molten silicon as a function of oxygen partial pressure of ambient atmosphere P_{O_2}.

tion, SIMS (secondary ion mass spectroscopy) analysis showed oxygen concentration of less than 2×10^{16} atoms cm^{-3}, i.e. the detection limit of the SIMS analysis for oxygen. The measured oxygen concentration was lower than that thermodynamically calculated. This is also the case for the specimens used for surface tension measurements by the levitation method. On the contrary, a silicon specimen employed for temperature oscillation experiment in microgravity showed oxygen concentration of 7×10^{17} atoms cm^{-3} (SIMS analysis). The origin of oxygen incorporation in the present experiment is the dissolution of the silica sheath of the thermocouples. An absolute value of temperature coefficient one order of magnitude smaller than that measured in the present work has been reported (-0.05 mN m^{-1} K^{-1}) (Popel' et $al.$ 1970) and was attributed to oxygen contamination (Keene 1987). Therefore, we cannot yet appropriately estimate the temperature coefficient of surface tension for the specimen in the microgravity experiment.

(c) Future works

Through the present work, it was made clear, as shown in figure 11, that the temperature coefficient of surface tension for molten silicon is sensitive to the oxygen partial pressure of the ambient atmosphere. This suggests that the Marangoni flow should also be sensitive to the oxygen partial pressure of ambient atmosphere, i.e. melt oxygen concentration and adsorbed oxygen on the melt surface. Therefore, we deduce that the mode of the Marangoni flow of molten silicon is quite different between that for Czochralski-growth and that for floating-zone-growth systems. For the Czochralski case, oxygen concentration is assumed to be high around the silica crucible wall which supplies oxygen, while it is very low for the floating-zone case.

The Marangoni flow in the liquid column corresponds to the floating-zone system, whereas the Marangoni flow on the flat surface corresponds to that in the Czochralski case. Although the Marangoni flow for the liquid column has been made clear to some

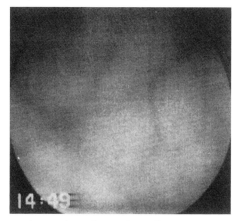

Figure 12. Network pattern observed at the surface of molten silicon (Yamagishi & Fusegawa 1990).

extent, nothing has been explained for that in the Czochralski case. Therefore, a study on the Marangoni flow for the flat surface melt should be promoted. As shown in figure 12, the network structure or the spoke pattern observed on the melt surface could be related to the Marangoni–Bénard instability (Yamagishi & Fusegawa 1990; Yi *et al.* 1994).

This study was supported by the Joint Research Program of the Japan Space Utilization Promotion Center under the direction of the National Space Development Agency of Japan.

References

Cröll, A., Müller-Sebert, W. & Nitsche, R. 1989 *Mat. Res. Bull.* **24**, 995.

Eyer, A., Leiste, H. & Nitsche, R. 1985 *J. Crystal Growth* **71**, 173.

Han, J. H., Sun, Z. W., Dai, L. R., Xie, J. C. & Hu, W. R. 1996 *J. Crystal Growth* **169**, 129.

Hibiya, T., Nakamura, S., Kakimoto, K., Imaishi, N., Nishizawa, S., Hirata, A., Mukai, K., Matsui, K., Yokota, T., Yoda, S. & Nakamura, T. 1996 *Proc. 2nd Euro. Symp. Fluids in Space, 22–26/4/1996, Naples*, p.231.

Hirata, H. & Hoshikawa, K. 1990 *J. Crystal Growth* **106**, 657.

Imaishi, N. & Yasuhiro, S. 1997 *Proc. 10th Int. Symp. Transport Phenomena, 1–3/12/1997, Kyoto, Japan.*

Kakimoto, K., Eguchi, M., Watanabe, H. & Hibiya, T. 1988 *J. Crystal Growth* **88**, 365.

Keene, B. J. 1987 *Surf. Interf. Analysis* **10**, 367.

Levenstam, M. & Amberg, G. 1995 *J. Fluid Mech.* **297**, 357.

Nakamura, S., Hibiya, T., Kakimoto, K., Imaishi, N., Nishizawa, S., Mukai, K., Yoda, S. & Morita, T. S. 1998 *J. Crystal Growth* (In the press.)

Niu, Z.-G., Mukai, K., Shiraishi, Y., Hibiya, T., Kakimoto, K. & Koyama, M. 1998 *Proc. Joint 10th European and 6th Russian Symp. Physical Sciences in Microgravity.* (In the press.)

Popel', S. I., Shergin, L. M. & Tsarevskii, B. V. 1970 *Russ J. Phys. Chem.* **44**, 144.

Przyborowski, M., Hibiya, T., Eguchi, M. & Egry, I. 1995 *J. Crystal Growth* **151**, 60.

Schwabe, D., Scharmann, A., Preisser, F. & Oeder, R. 1978 *J. Crystal Growth* **43**, 305.

Wanschura, M., Shevtsova, V. M., Kuhlmann, H. C. & Rath, H. J. 1995 *Phys. Fluids* **7**, 912.

Yamagishi, H. & Fusegawa, I. 1990 *J. Jap. Ass. Crystal Growth* **17**, 304.

Yi, K.-W., Kakimoto, K., Eguchi, M., Watanabe, M., Shyo, T. & Hibiya, T. 1994 *J. Crystal Growth* **144**, 20.

Marangoni effects in welding

By K. C. Mills, B. J. Keene, R. F. Brooks and A. Shirali

*Centre for Materials, Measurement and Technology, National Physical Laboratory,
Teddington, Middlesex TW11 0LW, UK*

The problem of *variable weld penetration* or 'cast-to-cast' variation in GTA/TIG welding is discussed. It is shown that for normal GTA/TIG welding conditions the Heiple–Roper theory is valid, i.e. that weld penetration is controlled by the fluid flow in the weld pool which, in turn, is controlled by the direction and magnitude of the thermocapillary forces. For most steels the direction and magnitude of these forces are determined by the sulphur content, since the temperature coefficient of surface tension $(d\gamma/dT)$ is negative when S $<$ 30 ppm and this leads to a radially outward flow and poor penetration whereas a steel with S $>$ 60 ppm has a positive $(d\gamma/dT)$ which produces a radially inward flow giving good weld penetration. Thermocapillary forces were shown to play a part in the problems of 'off-centre welding', 'porosity' and 'arc wander' in GTA/TIG welding and in the surface rippling of welds.

Keywords: TIG/GTA welding; penetration; thermocapillary forces; sulphur content

1. Introduction

The first example where Marangoni forces have been proposed to be involved in welding problems was in the case of *'cast-to-cast' variations* or *variable weld penetration.*

The problem of *'cast-to-cast'* variations in weld penetration produced during autogenous tungsten inert gas (TIG)† welding of stainless and ferritic steels was first noted in the 1960s. The problem is particularly severe in robotic processes requiring thousands of repetitive welds where it is customary to establish the welding parameters which promote deep penetration joints. However, it has been found (as can be seen in figure 1) that certain batches of steel produced welds with much lower weld penetration than the norm, despite fully meeting the material specifications. Welds can be partial-penetration welds (figures 1a, b) or full-penetration welds (figure 1c). It is customary to express penetration in partial welds by the depth (D)/width (W) ratio.

There have been several attempts to establish a correlation between 'cast-to-cast' variations and systematic variations in the concentrations of specific minor or impurity elements in the metal. However, where such a relationship could be identified it was noted that any such variations in the element concentrations were very small. Thus, any theory proposed to account for variable weld penetration must explain why such small differences in chemical composition can have such a large effect on 'weldability'. Several theories have been proposed in which it was suggested that small differences in minor element concentrations in the steel produced changes in

---

† Sometimes known as gas tungsten arc (GTA) welding in which an arc is struck between a cathode and the workpiece (anode).

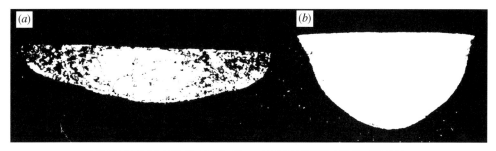

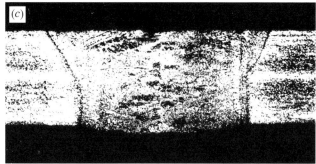

Figure 1. Comparison of the cross section of TIG weld fusion zones in (a) shallow- and (b) deep-penetration welds in stainless steel (×10) and (c) full-penetration weld, this case in penetration defined by the ratio of widths of back and front welds (W_b/W_f).

(a) the arc characteristics (Glickstein *et al.* 1977; Savage *et al.* 1977) and (b) the surface properties of the weld pool by affecting either the interfacial energies (Roper & Olsen 1978) or the fluid flow motion in the weld pool (Heiple & Roper 1982). However, it has also been found that variable weld penetration occurred in non-arc processes such as laser and electron beam welding (Robinson *et al.* 1982; Heiple *et al.* 1983) where there are no arc effects. Consequently, cast-to-cast variation could not be explained solely on the basis of changes in the arc characteristics and thus attention has been focused mostly on changes in the surface properties of the melt.

As can be seen from figure 2, small differences in the concentrations of surface-active elements, such as sulphur (Gupt *et al.* 1972/73) and oxygen (Gupt *et al.* 1976), cause substantial changes in the surface tension (γ) of iron and other elements. Friedman (1978) developed a model of the weld pool in which it was proposed that the surface-tension forces operating in the pool opposed the combined effects of gravity and arc pressure; thus a high surface tension would lead to poor weld penetration. Other theories have focused on the effect of the surface tension on the fluid flow in the weld pool. Ishizaki (1965) suggested that the surface-tension gradient ($d\gamma/dT$) across the pool could affect the convective flow in the weld pool. Heiple & Roper (1982) developed this theory, and postulated that variable weld penetration is a result of differences in the fluid flow in the weld pool resulting from differences in both the direction and magnitude of thermocapillary forces, and that these were controlled by the concentrations of surface-active elements such as sulphur and oxygen in the metal.

Heiple & Roper (1982) also pointed out that when the sulphur or oxygen concentration exceeded a certain critical value (around 50 ppm), the temperature coefficient of surface tension ($d\gamma/dT$) changed from a negative to a positive value (figure 3). They suggested that since a large temperature gradient exists between the centre and

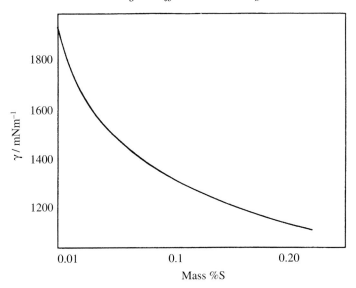

Figure 2. Variation of surface tension with sulphur content (Gupt *et al.* 1972/73).

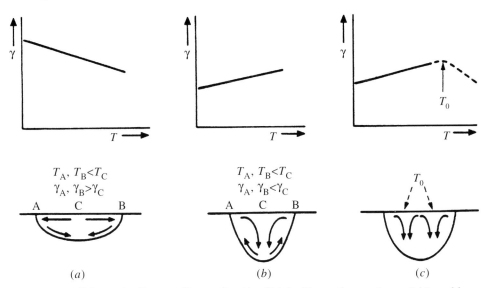

Figure 3. Schematic diagram illustrating the Heiple–Roper theory for variable weld penetration (Heiple & Roper 1982).

the edges of the weld pool (of the order of 500 K mm^{-1}), a large surface-tension (γ) gradient will be produced across the surface. The resulting Marangoni flow will occur from a region of low γ to a region of high γ. These surface flows subsequently trigger circulation flows in the molten weld pool, as shown in figure 3. For most pure metals, including iron and steels with low O and S contents, the surface tension decreases with increasing temperature, which results in a negative surface-tension-temperature coefficient ($\mathrm{d}\gamma/\mathrm{d}T$) (figure 3*a*). In this case, the surface tension will be greatest in the cooler regions at the edge of the weld pool and this induces a radially outward surface flow which carries hot metal to the edge of the pool where the consequent melt-back results in a wide shallow weld. In contrast to this, in Fe-based melts with

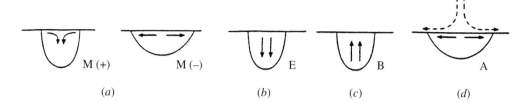

Figure 4. (a) Thermocapillary (Marangoni) forces $M(+)$ or $M(-)$; (b) electromagnetic (Lorentz) forces E, resulting from interaction of current; (c) buoyancy forces B, resulting from density differences caused by temperature gradients; (d) aerodynamic drag forces A, caused by passage of plasma over surface.

S (or O) > 60 ppm, $(\mathrm{d}\gamma/\mathrm{d}T)$ will be positive (figure 3b) and thus the surface tension is greatest in the high-temperature region at the centre of the pool and this induces a radially inward flow. This, in turn, produces a downward flow in the centre of the weld pool (figure 3b) which transfers hot metal to the bottom of the pool where melt-back of the metal results in a deep and narrow pool.

Keene et al. (1982) have pointed out that systems which exhibit a positive $(\mathrm{d}\gamma/\mathrm{d}T)$ must go through a maximum at some temperature and thus produce a complex flow similar to that shown in figure 3c.

2. Forces affecting the fluid flow in the weld pool

The Heiple–Roper theory makes two assumptions: (i) that the heat transfer in the weld pool is controlled by the fluid flow in the pool and not the heat conduction in the workpiece; and (ii) that the fluid flow is dominated by the thermocapillary forces.

However, there are several other fluid flow mechanisms operating in the weld pool, namely electromagnetic (or Lorentz), aerodynamic drag and buoyancy forces (figure 4). Under certain welding conditions these forces can have a significant effect on the fluid flow in the weld pool.

(a) Marangoni forces

These are, for the most part, thermocapillary forces but diffusocapillary forces can arise when welding steels with different sulphur contents (see § 7 a). The direction of the thermocapillary flow is determined by the concentration of O or S in the alloy.

The strength of the thermocapillary flow is determined by the non-dimensional Marangoni number (Ma) defined in equation (2.1) where $(\mathrm{d}T/\mathrm{d}x)$ is the temperature gradient, η is the viscosity, a is the thermal diffusivity and L is the characteristic length:

$$Ma = \frac{\mathrm{d}\gamma}{\mathrm{d}T}\frac{\mathrm{d}T}{\mathrm{d}x}\frac{L^2}{\eta a}. \tag{2.1}$$

(b) Electromagnetic or Lorentz forces

The Lorentz forces are caused by the interaction of the induced magnetic field and the current carried by a conductor. The welding current induces a magnetic field around the conductor and the Lorentz force acts inwards and downwards in the weld pool (figure 4b).

(c) *Buoyancy forces*

Buoyancy forces are caused by the density differences due to temperature gradients in the weld pool and result in an upward flow (figure 4c). However, it has been shown that buoyancy forces are generally very small in relation to the other forces in weld pools of less than 10 mm depth.

(d) *Aerodynamic drag forces*

These forces are produced by the action of the arc plasma flowing over the surface of the weld pool, which induce an outward flow along the surface of the pool (figure 4d).

However, the fluid flow in the weld pool is exceedingly complex since, thermo-capillary, Lorentz, aerodynamic and buoyancy forces can all influence the flow. The situation is further complicated by (i) the front-to-back flow resulting from the relative motion of the workpiece to that of the electrode which is particularly important at high welding speeds and (ii) the 'spin' developed by the liquid metal under conditions of radially inward flow, which tends to reduce the magnitude of the radially inward flow (Lancaster 1987).

The mathematical modelling of the relative strengths of the four forces affecting the fluid flow has become a subject of great interest in recent years and about twenty models have been reported (Mills & Keene 1990). Virtually all these models predict that the Marangoni forces are predominant under normal welding conditions and have a decisive effect on the weld profile (Oreper & Szekely 1984; Kou & Sun 1985).

3. Relation between penetration and surface tension

As mentioned previously, penetration is usually expressed by the ratios of the (depth/width) (D/W) and the (back/front) widths (W_b/W_f) for partial- and full-penetration welds, respectively; the latter parameter is subject to welding characteristics and is a less satisfactory measure than (D/W).

The link between weld penetration and surface tension of the alloy has been demonstrated by Mills *et al.* (1984) who used the levitated drop method to measure the surface tension of casts with good and bad penetration, i.e. high and low (D/W) ratios, respectively. A typical example is shown in figure 5a. It was found that:

(i) good weld penetration correlated with low values of surface tension (γ) and positive values of $(d\gamma/dT)$ as found in steels with high-sulphur (HS) contents;

(ii) poor weld penetration correlated with high values of γ and negative values of $(d\gamma/dT)$ as found in steels with LS contents.

Mills & Keene (1990) subsequently correlated $(d\gamma/dT)$ with the S contents of the steels and showed that the 'cross-over' point where $(d\gamma/dT) = 0$ occurred around 40 ppm S (figure 5b). Thus good weld penetration is obtained in steels with greater than 60 ppm S and poor penetration in casts with less than 30 ppm S.

Mathematical models have shown that fluid flow in the weld pool is complex, despite the fact that thermocapillary forces tend to be dominant. Nevertheless, on the basis of the Heiple–Roper theory, some correlation between (D/W) and $(d\gamma/dT)$ might be expected. The γ–T relationship was determined for three steels with different S contents and the (D/W) ratio was derived for partial welds carried out on 6 mm plates of these steels. It can be seen from figure 6 that there is a good correlation between (D/W) and $(d\gamma/dT)$.

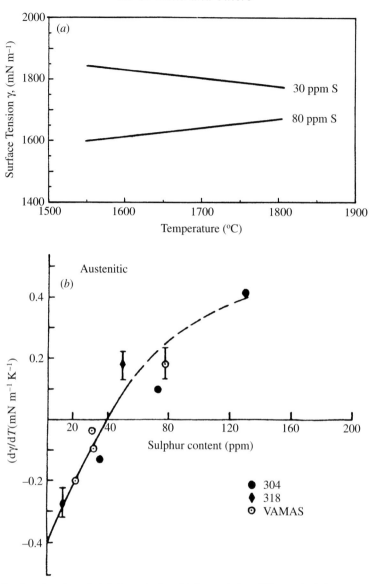

Figure 5. (*a*) Surface tension of stainless steels with low sulphur (30 ppm) giving poor penetration and high sulphur (80 ppm) and good penetration and (*b*) dependence of dγ/dT on sulphur content.

4. Effect of various elements on weld penetration

Elements can be classified into the following three classes.

(i) *Surface-active* elements (eg S, O, Se, Te), which affect the magnitude and direction of the fluid flow.

(ii) *Reactive* elements (eg Ca, Ce, Al), which react with the surface-active elements and thereby reduce the concentrations of soluble O and S.

(iii) *Neutral* elements which have little effect on the fluid flow in the weld pool.

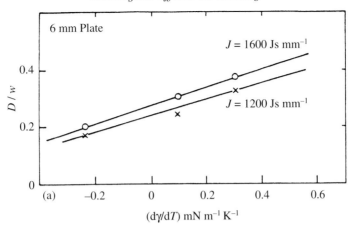

Figure 6. Penetration (D/W) as a function of ($d\gamma/dT$) for 6 mm-thick plates.

(*a*) *Surface-active elements*

(i) *Oxygen*

Although oxygen is almost as surface active as sulphur, Robinson & Gould (1987) have shown that it does not always have as great an effect on weld penetration as sulphur. It should be noted that it is the concentration of *soluble* O or S, denoted O, or S, which affects the surface tension since the *combined* oxygen (in the form of oxides) has little effect on the surface tension. However, it can be seen from figures 7a, b that the concentrations of scavenging elements, such as Al, in steel will hold the soluble O concentration below 10 ppm, whereas they do not have the same effect on the soluble S unless the steel contains large concentrations of Ca or Ce, which is rare. Thus, O $\ll$ O$_{total}$ and S $\simeq$ S$_{total}$ and therefore it is the sulphur and not the oxygen which has the greatest effect on the weld penetration.

(ii) *Sulphur*

Recent work on the effect of sulphur (Shirali & Mills 1993) on weld penetration was carried out on both high-sulphur (HS) and low-sulphur (LS) steels using various doping techniques; the results are shown in figure 8. It was found that for all the steels, an increase in S content produced an increase in the depth/width (D/W) ratio. These results are in essential agreement with Heiple–Roper theory and with results obtained by other investigators.

(*b*) *Reactive elements*

It can be seen from figure 7a that Al additions in excess of 20 ppm will react with the soluble oxygen present in the steel to form Al$_2$O$_3$ and thus reduce the soluble O concentration to a very low level. Under these conditions ($d\gamma/dT$) would become negative. Consequently, the thermocapillary forces would be expected to produce a radially outward surface flow resulting in a reduction of the (D/W) ratio of the weld. Calcium and cerium behave in a similar manner. Thus, providing there are sufficient amounts of these elements to react with the oxygen present, the soluble O concentration in steels will be below 5 ppm.

The Ca, Ce, La will react with soluble S to give their respective sulphides and reduce the S to very low levels, but this is not the case for Al (which is sited near to the Zr curve in figure 7b). Thus, providing the steel does not contain significant

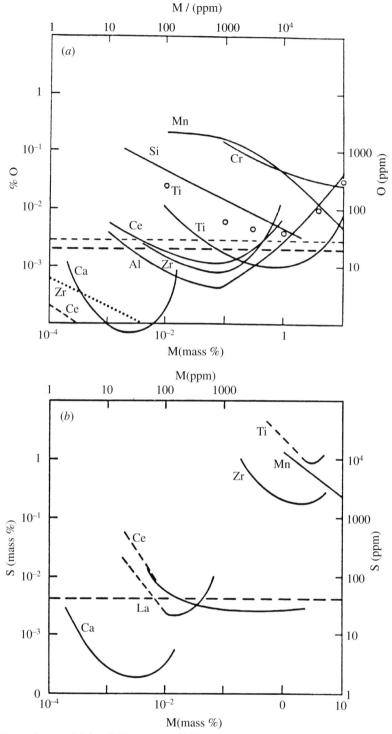

Figure 7. Dependence of (a) soluble oxygen (%O) on concentration of alloying element (M) in Fe–M–O systems at 1873 K; (b) soluble sulphur (%S) on concentration of alloying element (M) in Fe–M–S systems at 1873 K (Mills & Keene 1990).

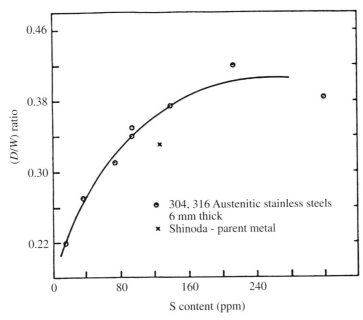

Figure 8. Effect of sulphur content on the (D/W) ratio of TIG welds (Shirali & Mills 1993).

levels of Ca, Ce, La, the soluble S̲ concentration will be only slightly less than the total S content. This is the reason why penetration can be readily correlated with S content but is less readily correlated with total O content.

5. Effects of 'slag spots' and oxide films

'Slag spots' are formed by the floating of non-metallic inclusions on the surface of the metal and they are produced by reactions of the metal with O and S. Pollard (1988) showed that they attract the arc and reduce the size of the anode root and thus increase the current density. In casts with S > 50 ppm the fluid flow will be radially inward and the slag will be sited in the centre of the pool; thus the resultant high current density will result in increased temperature gradients and, consequently, better penetration. However, in LS casts the flow will be radially outward and the slag spot will be swept to the periphery of the pool. The consequent attraction of the arc will result in deeper penetration at the edge of the pool, which leads to an erratic weld seam (see §7 *b*).

When the steel contains significant concentrations of Ca (greater than 20 ppm), the Ca forms an oxide film on the edge of the weld pool. These surface films tend to suppress surface flows and thus produce stagnant regions at the edges of the weld pool.

6. Effect of welding parameters on weld penetration

Burgardt & Heiple (1986) pointed out that since Marangoni forces are usually dominant in the weld pool, the effects of altering welding conditions can be explained in terms of what effects these changes would have on the temperature gradient (and hence the strength of Marangoni forces operating in the weld pool). Thus any change

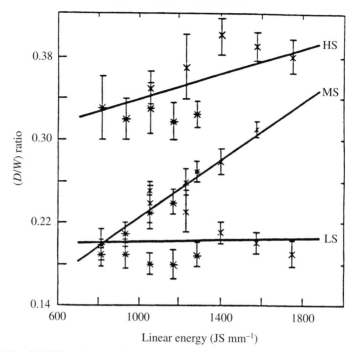

Figure 9. The (D/W) ratio as a function of the linear energy (error bars represent the standard deviation).

which brings about an increase in temperature gradient would cause increased penetration in HS casts and reduced penetration in LS casts. Although this proposition ignores the effect of welding parameters on the other forces operating in the weld pool, these workers did show that it could account for their observations.

Mills and Keene (1990) analysed the effect of changes in the welding parameters, such as arc length, welding speed (S_w) and current (I) and voltage (V), etc., on all four forces affecting fluid flow. The thermocapillary forces are affected by the temperature gradient (dT/dx), which is related to the power input (IV). The travel speed (S_w) affects the rate of heat input to the weld and it is this quantity which controls (dT/dx). Consequently, the effect of welding parameters on the Marangoni forces can best be studied from measurements of the heat input per unit length of weld. This parameter is referred to as the linear energy and is defined as (IV/S_w).

The linear energy allows the effect of current and welding speed on the Marangoni forces to be taken into account simultaneously but it should be noted that it does not account for either increased Lorentz forces with increasing current or the increased front-to-back motion in the weld pool at high welding speeds. The effect of increased linear energy on high- (HS) medium- (MS) and low-sulphur (LS) steels is shown in figure 9. It can be seen that increases in the linear energy resulted in increased penetration for HS and MS casts but have little effect on the LS casts. Thus these results are in essential agreement with Burgardt & Heiple's (1986) proposition that the effect of welding parameters on penetration can be explained in terms of their effect on the temperature gradients.

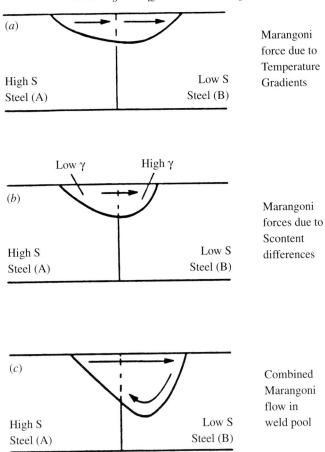

Figure 10. Schematic drawings showing the formation of a non-axisymmetric weld when welding steels with different sulphur contents: (*a*) Marangoni force due to temperature gradients; (*b*) Marangoni forces due to S content differences; (*c*) combined Marangoni flow in weld pool.

7. Other effects in TGI/GTA welding

(*a*) *Off-centre welding*

Tinkler *et al.* (1983) showed that when welding a 30 ppm S plate to a 90 ppm S plate, the resulting weld was off-centre and displaced towards the LS side. This can be accounted for if it is assumed that Marangoni forces dominate the fluid flow in the weld pool. It can be seen from figure 10 that the thermocapillary forces in the LS and HS will be from left to right and the diffusocapillary forces will also operate from left to right. Thus these surface flows will cause hot metal to be carried to the LS side and melt back off the steel will result in an asymmetric weld.

(*b*) *Arc wander*

It has been mentioned above that certain casts of steel exhibiting poor weld penetration tend to give an erratic weld seam. This is known as 'arc wander' and can be seen in figure 11. It frequently occurs with steels containing greater than 20 ppm Ca. It is caused by slag spots attracting the arc. For radially inward flows the slag spot will be centred in the centre of the pool and the attraction of the arc will result, sequentially, in a greater current density (because of the smaller anode root) and

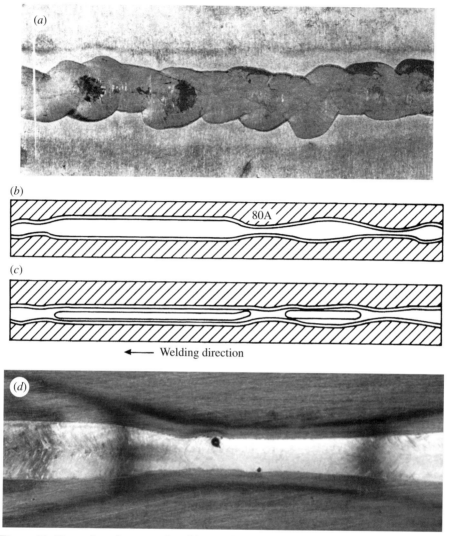

Figure 11. Examples of arc wander: (*b*) and (*c*) show the front and back faces of a full penetration weld; (*d*) is a magnified section of (*b*) showing presence of slag spots.

deeper penetration. With a radially outward flow the slag spot will be swept to the edge of the pool, since it attracts the arc, the hottest part of the pool will become the region close to slag spot and thus the position of the weld seam will change.

(*c*) *Porosity*

Poor weld penetration is often accompanied by porosity. Kou & Wang (1986) proposed that the direction of the flow in the weld pool could be responsible for the presence of pores in the weld. When the flow is radially inward the weld pool motion will assist the escape of bubbles away from the solidification front (figure 12*a*). In contrast, when the flow is radially outward the bubbles will tend to be swept towards the solidification front (figure 12*b*). It is obvious that the presence of a solid–slag film at the rear of the pool will tend to create a stagnant region which will assist the

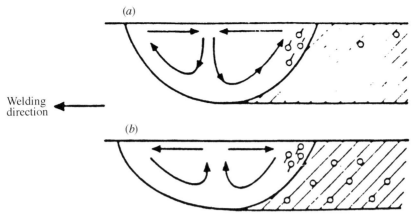

Figure 12. Influence of weld pool motion on porosity: (*a*) inward surface flow; (*b*) outward surface flow.

Figure 13. Predicted surface profiles of weld showing humping and undercutting: (*a*) inward surface flows with downward flow in centre; (*b*) outward surface flows with downward flow at periphery of weld pool.

entrapment of gas bubbles by the solidification front, thus it is not surprising that porosity problems are encountered when welding steels with high Ca levels.

(*d*) 'Humping' and 'undercutting'

'Humping' is the formation of a raised section in the centre of the weld and 'undercutting' is the depression at the edge of the weld (figure 13). 'Humping' and 'undercutting' were found to be prevalent for HS casts and when using high travel speeds. One characteristic of Marangoni flow is that the surface is raised in regions where the liquid is being driven downwards and depressed where the flow is upwards (figure 13). Thus for HS casts with a radially inward flow, it is obvious Marangoni flows can account for both humping and undercutting. However, on the basis of the flow patterns shown in figure 13 it is difficult to account for the undercutting in LS casts.

Gratzke *et al.* (1992) rejected the thermocapillary mechanism and proposed that humping and undercutting were caused by Rayleigh instability, i.e. the break-up of a liquid cylinder by the action of surface and gravity forces. They concluded that (i) the (width/length) ratio of the pool was the most important factor and (ii) the surface tension does not affect the onset of humping, only the kinetic behaviour which is a function of $(\rho/\gamma)^{1/2}$. The latter conclusion would seem to be inconsistent with the observation that it was prevalent in HS casts.

(*e*) Surface rippling

The surfaces of weld pools produced by LS casts tend to be flat and placid in contrast to the surfaces of HS casts which tend to be turbulent and agitated. The solidified welds of both HS and LS alloys exhibit regularly spaced fine ripples, possibly caused by the oscillation frequency of the arc or laser. However, for the HS casts there is a series of deeper ripples of longer wavelength superimposed on the fine ripple background. In full penetration welds the coarse ripples were seen on both free surfaces and pulsation of the laser would not be expected to affect the back surface. If the rippling is associated with the surface properties of the steel melt, then it must be related to a low surface tension and a positive $(d\gamma/dT)$. In the weld pool there are also thermal gradients in the direction perpendicular to the surface. Although these gradients do not produce any substantial fluid flow in the weld pool they can produce thermocapillary instabilities. These instabilities arise when a metal with a negative $(d\gamma/dT)$ is heated from below or a metal with positive $(d\gamma/dT)$ is heated from above and which give rise to capillary waves (Nemchinsky 1997). The thermocapillary forces acting parallel to the surface amplify these capillary waves and produce instabilities. Thus the surface rippling is probably due to the creation of thermocapillary instabilities. It also explains why they only occur in HS casts since they have positive $(d\gamma/dT)$ coefficients and welding is usually carried out by heating the upper surface.

8. Gas–metal arc welding (GMA)

This process is similar to GTA welding but a filler metal is used. Takasu & Toguri (this volume) have shown that there are four forces affecting the fluid motion in the pool when the molten filler metal drop hits the molten pool, namely: (i) a stirring force due to the momentum of the drop; (ii) a buoyancy force related to the density difference between the drop and the pool; (iii) a 'curvature' force related to the surface tension normal to the surface; and (iv) the Marangoni force related to the difference in surface tension of the drop and pool. Takasu & Toguri (this volume) showed that when (a) $\gamma_{drop} > \gamma_{pool}$ the droplet penetrated into the pool and (b) $\gamma_{drop} < \gamma_{pool}$ the drop will spread out over the surface.

9. Conclusions

(i) Variable weld penetration in TIG/GTA, laser and electron beam welding is caused by the direction and the magnitude of the fluid flow in the weld pool.

(ii) For normal welding conditions the fluid flow in the weld pool is controlled by the thermocapillary forces.

(iii) In most steels the direction of thermocapillary flow is determined by the sulphur content of the steel.

(iv) Other associated welding problems such as 'arc wander' and 'porosity' are also affected by the direction of fluid flow in the weld pool.

(v) The surface properties of the alloy may also be responsible for 'humping' and 'undercutting' and for the surface rippling of the weld.

References

Burgardt, P. & Heiple, C. A. 1986 Interaction between impurities and welding parameters in determining GTA weld shape. *Welding J.* **65**, 150s–155s.

Friedman, E. F. 1978 Analysis of weld puddle distortion and its effect on penetration. *Welding J.* **57**, 161s–166s.

Glickstein, S. S. & Yeniscavich, W. 1977 A review of minor element effects on the welding arc and weld penetration. *Welding Res. Council Bull.*, May 226.

Gratzke, U., Kapadia, P. D., Dowden, J., Kroos, J. & Simon, G. 1992 Theoretical approach to the humping phenomenon in welding processes. *J. Phys.* D **25**, 1640–1647.

Gupt, K. M., Yavoisky, N. I., Vishkaryov, A. F. & Bliznakov, S. A. 1972/73 *Met. Eng. Ind. Inst. Technol., Bombay* **1972/73**, 39–45.

Gupt, K. M., Yavoisky, N. I., Vishkaryov, A. F. & Bliznakov, S. A. 1976 *Trans. Ind. Inst. Metals* **29**, 286–291.

Heiple, C. A. & Roper, J. R. 1982 Mechanism for minor element effect on GTA fusion zone geometry. *Welding J.* **61**, 975s.

Heiple, C. A., Roper, J. R., Stagner, R. T. & Aden, R. J. 1983 Surface active elements effects on shape of GTA, laser and electron beam welds. *Welding J.* **62**, 72s.

Ishizaki, K. 1965 Interfacial tension theory of arc welding phenomena: formation of welding bead. *J. Jap. Weld. Soc.* **34**, 146.

Keene, B. J., Mills, K. C., Bryant, J. W. & Hondros, E. D. 1982 Effect of surface active elements on the surface tension of iron. *Can. Metall. Q.* **21**, 393.

Kou, S. & Sun, D. K. 1985 Fluid flow and weld penetration in stationary arc welds. *Metall. Trans.* A **16**, 203.

Kou, S. & Wang, Y. H. 1986 Weld pool convection and its effect. *Welding J.* **65**, 633.

Lancaster, J. F. 1987 The stability of meridionial liquid flow induced by a gradient of surface tension or by electromagnetic forces. *Int. Inst. Weld.*, doc. 212-682-87.

Mills, K. C. & Keene, B. J. 1990 Factors affecting variable weld penetration. *Int. Mater. Rev.* **35**, 185–316.

Mills, K. C., Keene, B. J., Brooks, R. F. & Olusanya, A. 1984 The surface tensions of 304 and 316 stainless steels and their effect on weld penetration. *Proc. Centenary Conf. Metallurgy Dept, University of Strathclyde, Glasgow, June 1984*, paper R.

Nemchinsky, V. A. 1997 The role of thermocapillary instability in heat transfer in a liquid metal pool. *Int. J. Heat Mass Transfer* **40**, 881–891.

Oreper, G. M. & Szekely, J. 1984 Heat and fluid flow phenomena in wled pools. *J. Fluid Mech.* **147**, 55.

Pollard, B. 1988 The effect of minor elements on the welding characteristics of stainless steel. *Welding J.* **67**, 202s–213s.

Robinson, J. L. & Gould, T. G. 1989 Effects of composition and physical properties on GTA weld penetration of austenitic stainless and low alloy steels. *Proc. ASTM & Int. Conf. on Trends in Welding Research, Gatlinburg TN, USA, 14–18/5/1989*.

Robinson, J. L., de Rosa, S. & Hutt, G. A. 1982 *Proc. Int. Conf. on the Effects of Residual Impurity and Microalloying Elements on Weldability and Weld Properties, London*, paper 41. Abington: The Welding Institute.

Roper, J. R. & Olsen, D. L. 1978 Capillarity effects in GTA weld penetration of 21-6-9 stainless steel. *Welding J.* **57**, 104s–107s.

Savage, W. F., Nippes, E. F. & Goodwin, G. M. 1977 Effect of minor elements on fusion zone dimensions of Inconel 600. *Welding J.* **56**, 126s–132s.

Shirali, A. A. & Mills, K. C. 1993 The effect of welding parameters on penetration in GTA welds. *Welding J.* **72**, 347s–352s.

Tinkler, M. J., Grant, I., Mizuno, G. & Gluck, C. 1983 *Proc. Int. Conf. on the Effects of Residual Impurities and Microalloying Elements on Weldability and Weld Properties, London*, paper 29. Abington: The Welding Institute.

Interfacially driven mass transport in joining and coating technologies

By David L. Olson and Glen R. Edwards

Center for Welding, Joining and Coating Research, Colorado School of Mines, Golden, CO 80401-1887, USA

The effects of interfacially driven or affected mass transport on (i) hot cracking of alloy weld metal, (ii) the role of interfacial chemical reactions at the leading edge of spreading braze material and (iii) the influence of interfacial tension gradients on abnormal grain growth in thin films are addressed. Hot cracking behaviour will be correlated to interfacial tension gradients in the interdendritic regions of the weld metal. Specific chemical reactions at the interface of the liquid metal braze on a ceramic substrate have been proposed to alter the interfacial force balances causing spreading phenomena. The potential role of interfacial tension gradients on abnormal grain growth in evolution of thin-film microstructures will be discussed.

The influence of Marangoni weld pool stirring and its effect on weld bead morphology has been thoroughly addressed in the literature. This paper will explore other areas in material processing where interfacially driven or affected mass transport can have an effect. First, we will discuss compositional and thermal gradient effects on hot cracking of alloy weld metal. The second area to be discussed will be the role of an interfacial tension gradient at the leading edge of spreading braze material during reaction brazing. The final area to be presented will be the influence of interfacial gradients on abnormal grain growth in thin films.

Keywords: solidification cracking; hot cracking; interdendritic flow; spreading of braze; thin-film growth; abnormal film growth

1. Influence on hot cracking

Hot cracking in weld deposits occurs due to the presence of low melting liquid films that allow boundaries to separate when thermal and shrinkage stresses develop during solidification and cooling (Borland 1960; Arata *et al.* 1976; Clyne & Davies 1979). Here, Fuerer's (1977) model for hot cracking will be first introduced, then combined with the concept of interfacially driven fluid flow to suggest some new insight into the mechanism of interdendritic cracking proposed by Holt *et al.* (1992).

Fuerer's (1977) hot cracking model makes the following assumptions.

(1) During plane front or cellular solidification, shrinkage will occur which is completely fed by residual liquid ahead of the interface.

(2) During dendritic solidification feeding becomes more difficult.

(3) Hot tearing is considered not possible if the rate of feeding (ROF) is larger or equal to the rate of shrinkage (ROS).

(4) Hot tearing becomes possible if the rate of shrinkage exceeds the maximum possible flow rate of feeding.

ROF is described by the following equation:

$$\text{ROF} = \left(\frac{\partial \ln V}{\partial t}\right)_{\text{feeding}} = \frac{g_L^2 d^2 P_s}{24\pi c^3 \eta L^2},\tag{1.1}$$

where

$$P_s = P_0 + P_c + P_m;\tag{1.2}$$

P_s is the effective feeding pressure, P_0 is the atmospheric pressure, P_c is capillary pressure, P_m is the metallostatic pressure head, g_L is the volume of liquid in the dendrite network, c is the tortuosity factor, η is the viscosity of the liquid phase, L is the size of the mushy zone, d is the secondary dendrite arm spacing, V is volume, and t is time.

ROS is described by the following equation:

$$\text{ROS} = \left(\frac{\partial \ln V}{\partial t}\right)_{\text{shrinkage}} = \frac{(\rho_0 - \rho_s + akC_L)\varepsilon g_L^{(2-k)}}{\rho(1-k)m_L c_0},\tag{1.3}$$

where

$$\rho = \rho_L g_L + \rho_s(1 - g_L);\tag{1.4}$$

ρ is the average density, ρ_L is the density of the liquid phase, ρ_0 is the density of liquid at the melting point, ρ_s is the density of the solid phase, a is the composition coefficient of liquid density, C_L is the composition of the liquid at the solid–liquid interface, C_0 is the alloy composition, ε is the average cooling rate during solidification, k is the equilibrium partitioning coefficient, and m_L is the slope of the liquidus line.

Figure 1 shows ROF and ROS for an alloy where the temperature (T) is above the eutectic melting temperature (T_E) of this alloy. Hot tears will not form as long as ROF is greater than ROS. It becomes important therefore, to have the intersection of the curves (i.e. T_i) at a point below the eutectic melting temperature if cracking is to be avoided.

Fluid flow driven by interfacial tension effects will be incorporated into Fuerer's (1977) hot cracking model (Holt *et al.* 1992; Cross *et al.* 1990). The force (F) acting on a surface–interfacial element as a consequence of a surface tension gradient can be described by the following equation:

$$F = -\frac{\partial U}{\partial x} = \frac{\partial \gamma}{\partial x} = \frac{\partial \gamma}{\partial T}\left(\frac{\partial T}{\partial x}\right) + \frac{\partial \gamma}{\partial C}\left(\frac{\partial C}{\partial x}\right),\tag{1.5}$$

where U is the potential energy which, by definition, is the negative of the Helmholtz free energy. Thus, $\partial U/\partial x$ is the negative of $\partial \gamma/\partial x$. Equation (1.5) shows that there will be a force acting on a surface–interfacial element whenever there is a temperature gradient or concentration gradient along that interface, provided $\partial \gamma/\partial T$ and $\partial \gamma/\partial C \neq 0$.

The fluid flow in the weld pool is driven by a surface tension gradient which is very sensitive to surface-active elements. Surface-active elements can cause reversal in the sign of the temperature dependence of the surface tension, and thus promote reversal of the fluid flow in the weld pool. This concept will be applied to fluid flows along the solid–liquid interface during dendritic solidification.

One of the characteristics of dendritic growth is the existence of a temperature gradient along the dendrite axis, with high temperatures at the dendrite tip (liquidus),

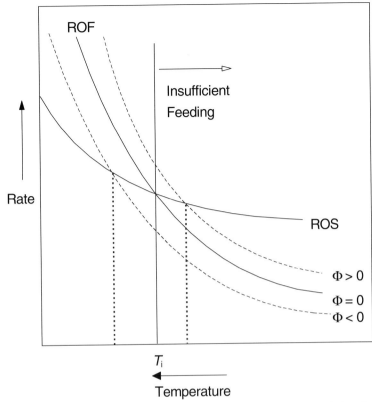

Figure 1. Hypothetical comparison of rate of feeding (ROF) and rate of shrinkage (ROS) as a function of temperature (T) in dendritic solidification; Φ is the pressure resulting from the surface tension gradient (Holt *et al.* 1992).

and low temperatures at the dendrite root (eutectic). This temperature gradient, coupled with $\partial\gamma/\partial T$, will result in a force acting on the interfacial elements along the solid–liquid interface, which will result in a fluid flow along this interface. The direction in which the liquid will flow depends upon the sign of $\partial\gamma/\partial T$.

Cross *et al.* (1990) have combined the Fuerer model with fluid flow driven by interfacial tension. It follows that, apart from the effective feeding pressure P_s, there can also be a 'pressure' resulting from a surface tension gradient acting on the interdendritic fluid (Φ). Equation (1.1) can now be rewritten:

$$\text{ROF} = \frac{g_L^2 d^2 (P_s + \Phi)}{24\pi c^3 \eta L^2}. \tag{1.6}$$

The 'pressure' Φ will, depending on its sign, shift the curve for ROF as shown in figure 1. When, as a consequence of interfacial-tension-driven flow, fluid is forced into the mushy zone, Φ will have a positive value and will result in a lowering of T_i. When, as a consequence of interfacial-tension-driven flow, fluid is forced out of the mushy zone, Φ will have a negative value which will raise T_i, having a deleterious effect on hot cracking resistance. The influence of $\partial\gamma/\partial T$ on interdendritic fluid flow and hot cracking tendency will be shown to be affected by the quantity of surface–interfacial active elements in cases 1 and 2. The influence of solute (i.e. surface–interfacial-active element) redistribution along the dendrite axis will be shown in case 3. The influence

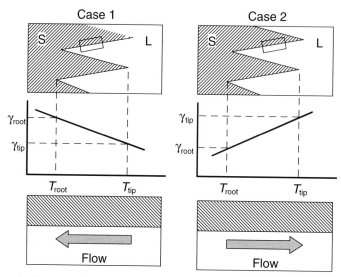

Figure 2. Schematic diagram showing interdendritic fluid flow influenced by the sign of $d\gamma/dT$: case 1, $d\gamma/dT < 0$ (low sulphur); case 2, $d\gamma/dT > 0$ (high sulphur) (Holt *et al.* 1992).

of temperature gradient on interdendritic fluid flow and hot cracking tendency will be shown in cases 4 and 5.

Even though a compositional gradient in a solid surface could not directly create Marangoni flow in an adjacent liquid, the consequence of the solute partitioning during solidification is to create a compositional gradient in the static liquid boundary layer. It is this liquid compositional gradient which causes Marangoni flow within the bulk interdendritic liquid. Such Marangoni flow inhibits the more normal backfilling, thus creating a liquid deficiency and an incipient hot crack.

(a) Influence of impurity concentration

From equation (1.5), the direction in which the force F works upon an interfacial element can be determined by the sign of $\partial\gamma/\partial T$. In the case of $\partial\gamma/\partial T < 0$ (case 1), fluid will flow from hot areas (γ_{low}) to cooler areas (γ_{high}). In the case of $\partial\gamma/\partial T > 0$ (case 2), fluid will flow from cooler areas (γ_{low}) to hot areas (γ_{high}). Figure 2 shows these two cases. It is seen from this figure that in case 1, fluid is forced into the mushy zone (i.e. $\Phi > 0$), resulting in a lowering of hot-cracking tendency, while in case 2 fluid is forced out of the mushy zone (i.e. $\Phi < 0$), resulting in an increase in hot-cracking tendency.

In practice the sign of $\partial\gamma/\partial T$ depends upon the absence or presence of surface–interfacial active elements. In the case of steels, sulphur has proven to be an important element determining the sign of $\partial\gamma/\partial T$ (Heiple & Roper 1982; Mills & Keene 1990; Burgardt & Campbell 1992). A low sulphur level results in $\partial\gamma/\partial T < 0$ and a high sulphur level results in $\partial\gamma/\partial T > 0$. Consequently, case 1 represents the situation of a low sulphur steel, while case 2 represents the situation of a high sulphur steel. Thus the model predicts a low hot-cracking tendency for low sulphur steels ($\Phi > 0$) and it predicts a high hot-cracking tendency for high sulphur steels ($\Phi < 0$).

During solidification there is also a redistribution of the surface active elements along the dendrite (case 3). This situation can also set up an interfacial tension gradient resulting in an assisted interdendritic fluid flow, as suggested in figure 3.

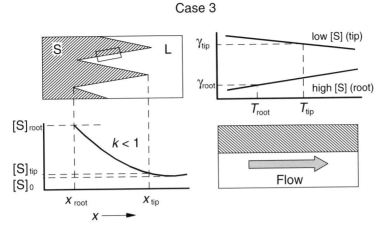

Figure 3. Schematic diagram showing interdendritic sulphur distribution and its influence on fluid flow (case 3) (Holt *et al.* 1992).

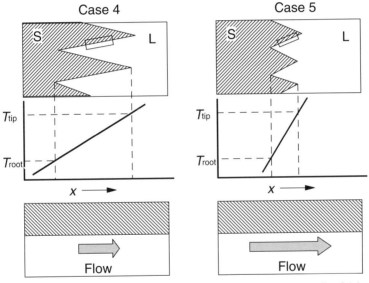

Figure 4. Schematic diagram showing influence of cooling rate on interdendritic fluid flow: case 4, small dT/dx (high heat input); case 5, large dT/dx (low heat input) (Holt *et al.* 1992).

(b) *Influence of temperature gradient*

The influence of the temperature gradient $\partial T/\partial x$ is now addressed. As can be seen from equation (1.5), the value of $\partial T/\partial x$ will influence the force acting on an interfacial element and will therefore influence the interfacial tension-driven fluid flow. In figure 4 this flow is described for a low (case 4) and a high (case 5) value of the temperature gradient and for $\partial\gamma/\partial T > 0$. During welding, a low cooling rate (high heat input) results in a low value of $\partial T/\partial x$, while a high cooling rate (low heat input) results in a high value of $\partial T/\partial x$. Case 4 therefore represents the situation during welding with a low cooling rate, while case 5 represents the situation during welding with a high cooling rate. As can be seen from figure 4, a small temperature gradient results in a relatively weak fluid flow while a large temperature gradient results in a relatively strong fluid flow. In both cases (note $\partial\gamma/\partial T > 0$) fluid is

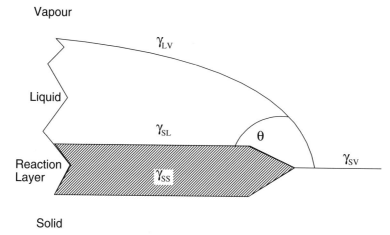

Figure 5. The formation of a reaction product between the reactive liquid brazing material and the solid substrate (Chidambaram *et al.* 1992).

forced out of the mushy zone and therefore $\Phi < 0$. Another factor influenced by the temperature gradient is the size of the mushy zone (L). A small temperature gradient produces a large mushy zone and a large temperature gradient creates a small mushy zone. As can be seen from equation (1.6), both Φ and L influence the rate of feeding (a large mushy zone has a strong detrimental effect on feeding property). Therefore, both factors have to be taken into account before the resulting hot cracking properties can be understood.

The model incorporates the effects of segregation and solute redistribution, which can be used to explain differences in hot-cracking tendencies between stainless steels solidifying as primary ferrite and those alloys solidifying as primary austenite. It also accounts for the effect of differences in cooling rate on hot-cracking susceptibility. It is understood that hot cracking is a complex phenomena involving many different factors in addition to the tendency for backfilling (Cross *et al.* 1990). Which of these factors has overriding influence on cracking susceptibility and the relative importance of interfacial tension in controlling hot crack feeding behaviour has yet to be determined.

2. Influence on spreading

The dynamics of a reactive alloy spreading on a ceramic can be studied as two distinct phenomena: (a) stage I, a reaction layer formation beneath the liquid drop; and (b) stage II, spreading of the liquid ahead of the original triple point.

Stage I. In reactive brazing, the reactive metal attacks the ceramic and forms a reaction product as suggested in figure 5. Chidambaram *et al.* (1992) has demonstrated that a surface thermodynamic criterion could be used to study the wetting and non-wetting metal–ceramic systems. Thermodynamics can only be used to predict wetting and identify the driving force in these systems. Further reaction and spreading of the liquid are entirely kinetic phenomena, and equilibrium thermodynamic will not sufficiently characterize these phenomena. Of interest to this paper is the region at the leading edge of the spreading liquid where the reaction product is not uniformly covering the substrate material, as suggested in figure 6.

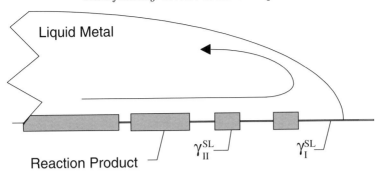

Figure 6. The partly covered reaction product region adjacent to the leading edge of the liquid brazing material. The partly covered reaction product region forms a gradient in liquid–solid interfacial energy and promotes spreading force in the liquid.

Stage II. The movement of the liquid ahead of the original triple point is described as spreading. In non-reactive systems, the driving force for spreading is the decrease in the total surface energy of the system given by the spreading coefficient (Heslot *et al.* 1989):

$$S = \gamma^{\mathrm{SL}} - \gamma^{\mathrm{SV}} - \gamma^{\mathrm{LV}}. \tag{2.1}$$

Spreading occurs when the spreading coefficient is positive. When S is negative, the liquid assumes an equilibrium contact angle. In non-reactive systems, spreading usually occurs after the liquid drop attains a zero degree contact angle. In reactive wetting, this behaviour is not observed; the liquid drop can assume an acute angle even when there is a driving force for spreading. The angle subtended by the drop is a manifestation of various kinetic events, and the angle can decrease only after the liquid drop has moved to an unreacted area.

As mentioned before, spreading in a reactive system is a more complex phenomenon than that described by equation (2.1). When the brazing alloy first comes into contact with the ceramic substrate, no reaction has occurred; hence, the drop does not yet wet the ceramic surface. The interfacial energy, $\gamma_{\mathrm{I}}^{\mathrm{SL}}$, in this situation is largely positive. The alloy then reacts with the ceramic according to stage I kinetics and causes the liquid drop to assume an acute angle, which is the situation where the reaction product has partly covered the surface. The liquid drop experiences two distinct interfacial energies from the reacted and unreacted fractions of the surface. If $\gamma_{\mathrm{II}}^{\mathrm{SL}}$ is the interfacial energy between the reaction product and the alloy, the resultant γ^{SL} can be calculated from a rule of mixtures approach:

$$\gamma^{\mathrm{SL}} = (1 - X)\gamma_{\mathrm{I}}^{\mathrm{SL}} + X\gamma_{\mathrm{II}}^{\mathrm{SL}}, \tag{2.2}$$

where X is the fraction covered at any given instant. This γ^{SL} is a function of coverage on the ceramic surface and, therefore, is a function of time, temperature, and composition of the reactive component of the brazing alloy. Notice that near the leading edge, a gradient in coverage by the reaction product occurs, and thus a gradient in the average interfacial tension is also present.

The observed spreading rate is governed by the balance between the driving force and the resisting force. The viscosity of the alloy would offer resistance to spreading of the drop. Using the analysis of Lopez *et al.* (1976) for a non-reactive drop, the viscous force is given by

$$F_v = \frac{\mu r^5}{\pi V t},\qquad(2.3)$$

where μ is the viscosity of the molten metal, r is the radius, t is the time and V is the volume of the drop. This term is a function of time and drop radius, but it is not affected by the extent of reaction at the ceramic–metal interface (assuming viscosity is not a function of the reactive metal content in the base alloy).

While the resistance force is the same for a reactive or non-reactive system, the driving force is significantly changed. For a non-reactive spreading system that assumes an instantaneous contact angle ϕ, Yin (1969) calculated the following driving force:

$$F_s\ (\text{non-reactive}) = 2\pi r S_\phi,\qquad(2.4)$$

where S_ϕ is the instantaneous spreading coefficient:

$$S_\phi = \gamma^{SL} - \gamma^{SV} - \gamma^{LV}\cos\phi.\qquad(2.5)$$

For this situation, the γs are not constants and $\cos\phi$ is determined by requiring the sessile drop to maintain a spherical cap as the drop spreads. In reactive systems, γ^{SL} in the equation for S_ϕ (equation (2.4)) is calculated using the rule of mixtures approach shown in equation (2.2). Therefore, the driving force for reactive spreading is

$$F_s\ (\text{reactive}) = 2\pi r[(1 - X)\gamma_I^{SL} + X\gamma_{II}^{SL} - \gamma^{SV} - \gamma^{LV}].\qquad(2.6)$$

For a non-reacting system, the resultant force is the difference in driving force and resistive force terms, and a steady-state spreading rate can be calculated (Lopez *et al.* 1976; Yin 1969). In a reactive system, the analysis is complicated by several factors and there is no longer a steady-state solution. Both the driving force and resistive force terms are functions of radius and time while the radius is also a function of time. Therefore, a nonlinear differential equation must be solved to obtain the spreading rate.

The variable which complicates the calculation of the spreading rate is the fraction of surface reacted (X) as a function of time and radius. Estimation of X is further complicated by the depletion of the reactive metal from the liquid metal droplet. As mentioned before, depletion occurs by both the formation of an oxide layer on the metal droplet at the liquid–vapor interface and by the thickening of the reaction layer at the substrate–liquid interface. Now considering the localized gradient in the interfacial tension near the leading edge, there is a second contributing term to the total driving force, F_{TS}:

$$F_{TS}\ (\text{reactive}) = 2\pi r[(1 - X)\gamma_I^{SL} + X\gamma_{II}^{SL} - \gamma^{SV} - \gamma^{LV}] + \pi r^2(\gamma_{II}^{SL} - \gamma_I^{SL})\frac{\partial X}{\partial r},\quad(2.7)$$

where the coverage, X, is a function of r. It is this second term which promotes the continual spreading of the liquid braze material on a ceramic substrate.

A complete understanding of a brazing system culminates in characterization of spreading (stage II). The theory presented further suggests that necessary information for the complete analysis includes the rate of coverage, the rate of interfacial phase formation (stage I kinetics), and the nature of the partial coverage of the reaction product in the region of the leading edge.

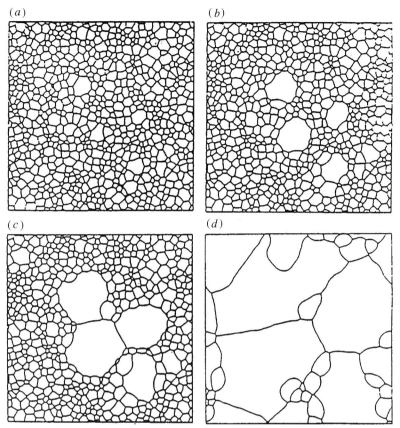

Figure 7. Evolution of the thin-film columnar structure progresses from the stagnation condition (*a*) to abnormal growth of a few grains (*b*), (*c*), to a final large grain structure (*d*) (Frost 1994).

3. Influence on growth of thin films

The evolution of thin-film microstructures progresses through a series of steps starting with nucleation, then growth to impingement, normal growth, and finally abnormal growth. Depending on the temperature of the substrate and the energy of the arriving atoms, various initial structural configurations will occur and with localized mass transport the structural defects are reduced and the grain growth will go from columnar to a larger equiaxed structure. The resulting thin-film structures were classified by Movchan & Demchishin (1960) and reclassified by Thornton (1988) as zones 1A, 1B, 2, T, and 3. This paper will consider interfacial tension gradients in the thin-film structures as a factor which promotes abnormal growth when the thin-film grains progress from a zone 2 to a zone 3 morphology.

After the grains have nucleated on the substrate and grown to impingement, grain boundary migration results in grains becoming large relative to the film thickness. The resulting grain boundaries become approximately perpendicular to the plane of the film and have a monomodal grain size distribution, as suggested in figure 7*a*. This arrangement of grain boundaries allows for two-dimensional grain growth modelling (Frost 1994).

Normal grain growth is modelled with the relationship for the velocity, v, given as

$$v = Mk\gamma_{\text{gb}}, \qquad (3.1)$$

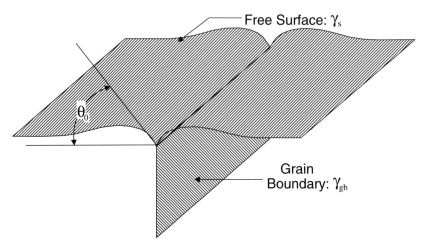

Figure 8. The morphology of a grain boundary groove between two columnar thin-film grains. The grain boundary grooves serve as pinning sites for thin-film grain growth (Mullins 1956).

where k is the curvature, M is the mobility and γ_{gb} is the grain boundary energy. It has been observed that normal grain growth often stagnates when the average grain diameter is two or three times the film thickness (Palmer *et al.* 1989; Frost *et al.* 1958). At the point of stagnation the grains are columnar and their boundaries completely traverse the thickness of the film (Frost 1994).

Mullins (1956) and Dunn (1966) has suggested that the stagnation of normal grain growth in the film is due to grain boundary grooving at the triple points where the grain boundary meets the free surface of the film, as illustrated in figure 8. In his model, surface diffusion redistributes the matter at the triple line so as to achieve equilibrium of the interfacial tensions. The angle at the bottom of the groove is determined by the force balance of the interfacial tensions. This angle is a measure of the force required to allow a grain boundary to escape from the groove. If the force is sufficient to pull the grain boundary to the angle Θ_0, the boundary will climb out of the groove.

If the grain growth proceeds beyond the stagnation of monomodal grain size distribution, abnormal growth occurs where a few grains are released to continue the grain growth (Mullins 1956; Thompson 1985; Thompson & Smith 1984; Wong *et al.* 1986). Figure 7 illustrates this abnormal growth behaviour. If the driving force is sufficient to pull the grain boundary to the angle Θ_c, the grain boundary can climb out of the groove. The force to promote escape or release of the grain boundaries may be the result of in-plane curvature which can pull the grain boundary to a position at which the intersection with the surface deviates from the perpendicular by

$$\Theta_c = \tfrac{1}{2}hk_{crit}, \qquad (3.2)$$

where h is the film thickness and the critical curvature for escape is k_{crit}, which is given by

$$k_{crit} = \frac{2\Theta_c}{h} \simeq \frac{\gamma_{gb}}{\gamma_s h}. \qquad (3.3)$$

Now consider that the thin film has grain boundaries with solute or contaminant compositional gradients which will alter the force balance suggested by Mullins (1956) in figure 9. The compositional gradients suggest an additional force term resulting

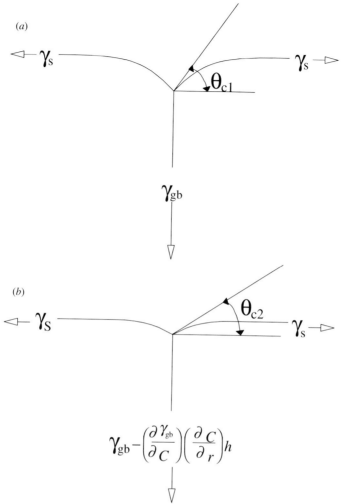

Figure 9. The reduction of the grain boundary tension due to extra force/length resulting from a compositional gradient along the grain boundary, which can be seen by comparing (a) with (b), to reduce the angle θ_c, which is a direct representation of the pinning force restraining grain growth.

in the following equation:

$$2\gamma_s \cos \Theta_c = \gamma_{gb} - \left(\frac{\partial \gamma_{gb}}{\partial C}\right)\left(\frac{\partial C}{\partial r}\right)h. \qquad (3.4)$$

If the grain boundary tension, γ_{gb}, is a function of solute or contaminant composition, then the critical angle to overgrow the grain boundary could be less. This situation would make abnormal growth easier to achieve, and would promote fewer but larger thin-film grains. According to Frost & Thompson (1988), Rolett *et al.* (1985) and Hillert (1965), once the grain boundary is released, two-dimensional growth models based on interfacial forces adequately describe the abnormal growth rate. The actual situation is more complex than presented, since there are two grain boundary grooves, as suggested in figure 10 (one grain boundary groove is on the free surface and the other groove is with the substrate).

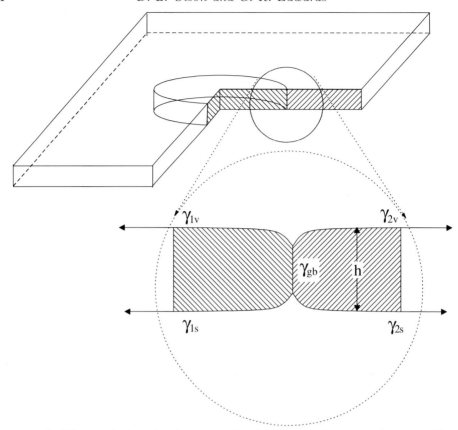

Figure 10. A difference in interfacial energies for the cylindrical grain and the neighbouring grain results in a driving force for grain growth after the grain boundary has been released (Frost 1994).

A suggested driving force for abnormal grain growth is the difference in surface energies for two neighbouring grains, $\Delta\gamma = \gamma_1 - \gamma_2$. The effect of this difference on the grain boundary migration rate is a function of the thin-film thickness, h. This concept is suggested in figure 10 (Frost 1994), where a circular grain in a thin film is shown experiencing the driving forces due to grain boundary capillarity.

By considering the driving force due to both the difference in the specific free surface energies between the inside and outside of the cylindrical boundary and the grain boundary capillarity, the total migration velocity, v, can be described by

$$v = v_{\text{gb}} + v_{\text{s}}, \tag{3.5}$$

where

$$v_{\text{gb}} = M\gamma_{\text{gb}}k \tag{3.6}$$

and

$$v_{\text{s}} = M\frac{2\Delta\gamma}{h}. \tag{3.7}$$

The $\Delta\gamma$ is assumed to be the result of the difference in surface energy due to crystal orientation between neighbouring grains. This situation is not very likely to be the primary factor since the thin-film columnar grains are growing in a preferred direction and thus exposing approximately the same crystal surface. A more likely

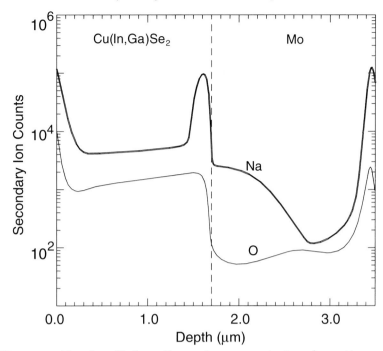

Figure 11. The compositional profile for sodium and oxygen content, surface active contaminants, across a Cu(In,Ga)Se₂–molybdenum thin film deposited on a soda glass. Notice the localized compositional gradients (after Scofield *et al.* 1994).

situation is a variation in surface contamination across the thin-film surface, grain to grain.

It is common to find compositional gradients in compound thin films, and these compositional gradients would be expected to alter the interfacial forces and thus the growth behaviour (Probst *et al.* 1996; Scofield *et al.* 1994; Kim & Thompson 1990). The influence of compositional gradients of sodium in copper–indium–gallium dis-elinide depositions have been shown to promote thin-film grain growth (Probst *et al.* 1996; Scofield *et al.* 1994). The source of the sodium is the soda glass substrate. This compositional gradient across a CIS–molybdenum deposit on a soda glass substrate can be seen in figure 11. Kim & Thompson (1990) have also reported the effect of dopants on surface-energy-driven secondary grain growth in silicon films.

4. Conclusion

The joining and coating processes produce material with significant gradients in composition, microstructure and properties. These gradients must be considered in achieving a mechanistic understanding of the microstructural evolution and the microstructure-property relationships of these processed materials.

The authors appreciate and acknowledge the research support of the Office of Naval Research.

References

Arata, Y., Matsuda, F., Nakata, K. & Sasaki, I. 1976 *Trans. JWRI* **5**, 53.
Borland, J. C. 1960 *Br. Welding J.* **7**, 508.

Burgardt, P. & Campbell, R. D. 1992 *Key Engng Mater.* **69/70**, 379–416.

Chidambaram, P. R., Edwards G. R. & Olson, D. L. 1992 *Met. Trans.* B **23**, 215.

Clyne, T. W. & Davies, G. J. 1979 *TMS Proc. Int. Conf. Solidification, Sheffield, UK*, p. 275.

Cross, C. E., Tack, W. T. & Loechel, J. L. W. 1990 *ASM Conf. Proc. On Weldability of Materials, Detroit*, p. 275.

Dunn, C. G. 1966 *Acta Metall.* **14**, 221–222.

Frost, H. J. 1994 *Mater. Character.* **32**, 257–273.

Frost, H. J. & Thompson, C. V. 1988 *J. Electron. Mater.* **17**, 447–458.

Frost, H. J., Thompson, C. V. & Walton, D. T. 1958 *Acta Metall.* **6**, 414–427.

Fuerer, U. 1977 *Proc. Int. Sym. Eng. Alloys, Delft, The Netherlands*, p. 131.

Heiple, C. R. & Roper, J. R. 1982 *Welding J.* **61**, 97.

Heslot, F., Fraysse, N. & Cazabat, A. M. 1989 *Nature* **338**, 640.

Hillert, M. 1965 *Acta Metall.* **13**, 227–238.

Holt, M., Cross, C. E. & Olson, D. L. 1992 *Scripta Metall. Mater.* **26**, 1119–1124.

Kim, H. J. & Thompson, C. V. 1990 *J. Appl. Phys.* **67**, 757–767.

Lopez, J., Miller, C. A. & Ruckenstein, E. 1976 *J. Colloid. Interf. Sci.* **56**, 460.

Mills, K. C. & Keene, B. J. 1990 *Int. Mater. Rev.* **35**, 185.

Movchan, B. A. & Demchishin, A. V. 1969 *Fiz. Met. Metall.* **28**, 653.

Mullins, W. W. 1956 *J. Appl. Phys.* **27**, 900–904.

Palmer, J. E., Thompson, C. V. & Smith, H. I. 1987 *J. Appl. Phys.* **62**, 2492.

Probst, V., Krag, F., Rimmasch, J., Riedl, W., Stetter, W., Harms, H. & Eibl, O. 1996 *Mat. Res. Soc. Symp. Proc. on Thin Films for Photovoltaics and Related Device Application*, vol. 426, p. 165.

Rollett, A. D., Srolovitz, D. J. & Anderson, M. P. 1985 *Acta. Metall.* **33**, 2233–2247.

Scofield, J. H., Asher, S., Albin, D., Tuttle, J., Contreas, M., Niles, D., Reedy, R., Tenant, A. & Noufi, R. 1994 *Proc. 1st WCPEC, 5–9/12/94, Honolulu, Hawaii*, pp. 1647–1667. IEEE.

Thornton, J. A. 1988 *A. Rev. Mater. Sci.* **7**, 239.

Thompson, C. V. 1985 *J. Appl. Phys.* **58**, 763–772.

Thompson, C. V. & Smith, H. I. 1984 *Appl. Phys. Lett.* **44**, 603–605.

Wong, C. C., Smith, H. I. & Thompson, C. V. 1986 *Appl. Phys. Lett.* **48**, 335.

Yin, T. P. 1969 *J. Phys. Chem.* **7**, 2413.

Discussion

R. C. COCHRANE (*School of Materials, University of Leeds, UK*). What is the scale at which the abnormal grain growth in thin films would be expected to occur? Is it at a scale of μm or tens of μm? (It could occur up to 0.5 mm or so!)

D. L. OLSON. It has been observed that abnormal grain growth initiates when the grain diameter in the plane of the film is 2–3 times the film thickness. Typical thin-film thicknesses are from 2–10 μm.

Observations of pulsating Marangoni phenomena during the local oxidation of deoxidized liquid steel

By G. R. Belton, T. J. Evans and L. Strezov

BHP Research, Newcastle Laboratories, PO Box 188,
Wallsend, NSW 2287, Australia

Studies have been made by means of a high-speed video technique of the surface of Al–Si–Mn deoxidized liquid steel at about 1600 °C during local reoxidation by gently impinging flows of CO_2-containing gases. Expanding 'necklaces' of alumina oxidation product have been observed to form at frequencies of up to about 5 Hz and to travel at velocities of up to 250 mm s^{-1} near to the gas inlet tube, depending on the experimental conditions. The pulsation frequency appears to be controlled by the rate of diffusion of aluminium to the surface. The effects of additions of the surfactants sulphur and selenium on the velocity of the moving front are shown to be consistent with the processes being surface tension driven.

Keywords: Marangoni flow; steel reoxidation; surface velocities; effect of surfactants; local oxidation; alumina formation

1. Introduction

Thermodynamic and kinetic phenomena at the gas–solid, gas–liquid, solid–liquid, liquid–liquid and liquid 1–liquid 2–solid interfaces are pervasive in iron and steel making. Olette (1993) expressed this succinctly in the title of his Yukawa Memorial Lecture, 'Surface phenomena: a cornerstone of iron and steel making processes', in which he discussed several important examples where interfacial behaviour is a determining factor. Studies of Marangoni effects in particular have been reviewed by Hammerschmid (1987) and, more recently, by Mukai (1992). The often profound effects of surface-active elements on the rates of interfacial reactions of gases with liquid iron and its alloys have also been reviewed recently by Belton (1993).

The present paper reports and discusses observations of concentration-driven Marangoni-induced pulsating flows which occur during the oxidation of aluminium-containing deoxidized steel by gently impinging flows of oxidizing gases under essentially isothermal conditions. The work was undertaken as part of broader studies of the phenomena which lead to disturbances of the meniscus at the liquid steel–gas–mould interface in the strip casting of steel. In particular, the studies are of initially manganese–silicon deoxidized steel with additions of aluminium. The results are also of possible interest in understanding the phenomena which lead to the blockage of submerged entry nozzles by agglomerated alumina particles.

2. Experimental details

About 135 kg of liquid steel was contained within a 300 mm internal diameter magnesia crucible in an induction furnace. By using this large mass, the temperature of the steel decreased only slowly from about 1600 to 1580 °C over a period of 4–5 minutes, when the induction power was switched off. The melt was sealed against the atmosphere by means of a refractory-coated steel plate and dense fibrous refractory gasket. The bulk atmosphere was provided by flowing a 5% H_2–Ar gas mixture through a tube in the steel plate, with provision of a restricted outlet. The oxidizing gas mixture was introduced through a vertical 5 mm bore alumina tube, with the tip held at about 5 mm above the liquid metal surface. In one series of experiments, pure CO was passed through a second tube which was similarly positioned at a lateral distance of 50 mm, centre to centre, from the first. In other experiments, the second tube was dispensed with. Temperature measurement was by means of an alumina-sheathed noble metal thermocouple inserted in the liquid metal, and the flow rates of the high purity gases were controlled by mass flow meters. Sampling of the melt for chemical analysis was through a sealable port in the cover plate.

Observation and recording of the events occurring on the surface of the liquid steel was by means of a high-speed video camera sighted through an optical flat cemented on a sight tube in the steel cover plate. Several optical filters were tried, but the most suitable was found to be a lens from a pair of arc-welding goggles. Reduction of the glare was also found to be assisted by making the refractory lining of the steel cover plate convex rather than flat.

The studies were carried out with melts which had been prepared for other experiments, and hence were manganese–silicon deoxidized low-carbon (0.001–0.07 wt% C) steels. Aluminium was added through the sampling port, the system closed, and the 5% H_2–Ar bulk atmosphere gas passed at a flow rate of about 5 l min^{-1}. The reaction gases were introduced through one or both of the 5 mm bore inlet tubes. After the initial alumina oxidation product had cleared from the surface, and following stabilization of the temperature at 1600 °C, the power to the induction furnace was turned off and observations begun. Video recording, at 25–500 frames per second, was typically begun after about a further 30 s. Observations were concluded when the temperature had decreased to about 1580 °C.

Quantitative measurements of pulsation frequencies and velocities were essentially manual, i.e. through repeated replaying of the video tapes at various speeds and counting, and by measurement of the distances moved in successive frames.

3. Observations and results

(a) *High aluminium-content melts with an oxidizing gas flow rate of about* 100 ml min^{-1}

The compositions of the melts for three successive sets of experiments were: 0.0190 ± 001 wt% Al, 0.071–0.074 wt% C, 0.039–0.042 wt% S, 0.32 wt% Mn and 0.048–0.057 wt% Si. The oxidizing gas had a p_{CO}/p_{CO_2} ratio of 5 and pure CO was passed down the second tube at a flow rate of about 25 ml min^{-1}.

The principal observations are summarized in the schematic, and somewhat idealized, figure 1. What appeared to be 'necklaces' of the oxidation product, presumed to be alumina, were generated at a regular frequency of 4–5 Hz. The average velocity of the expansion of the necklace was 120–180 mm s^{-1} at a distance of about 10 mm

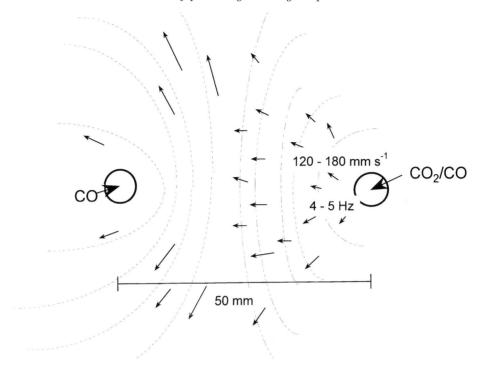

Figure 1. Schematic representation of the movement of 'necklaces' of alumina oxidation product in the experiments with the 0.019 wt% Al melt and a gas flow rate of 100 ml min^{-1}.

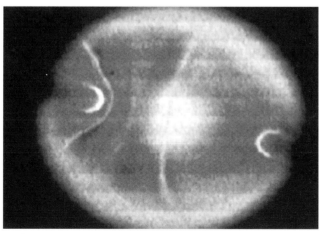

Figure 2. The expanding arc of the oxidation product, taken at a frame speed of 125 s^{-1}. The accumulation of oxidation product can be seen near the CO delivery tube on the left.

from the outer edge of the gas delivery tube. At a distance of 20–25 mm from the tube, the velocity was 60–80 mm s^{-1}. As the 'necklaces' approached to within about 10–15 mm of the CO gas delivery tube, motion towards the tube essentially stopped and particulate alumina moved towards the edges of the field of view. No pulsations were observed in the region of the CO delivery tube.

Figure 2 is from a single frame, taken at 125 s^{-1}. Although the resolution is poor, one expanding arc of the oxidation product is clear, as is the accumulation of the

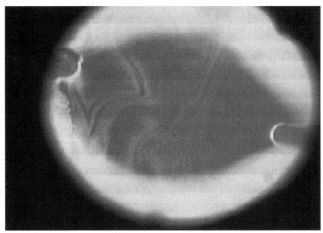

Figure 3. The accumulation of ribbons of oxidation product during an experiment with the 0.019 wt% Al melt during a period when a slow surface flow was occurring from left to right. The frame speed is 125 s^{-1}.

product near to the CO delivery tube on the left-hand side. The white rings under the gas inlet tubes are reflected light from the bases of the tubes. Figure 3 is taken from a separate experiment where the glare was reduced. Although distorted by a slow drift of the surface of the liquid iron from left to right, presumably driven by a temperature gradient, the accumulation of ribbons of oxidation product is reasonably clear.

(b) *Lower aluminium-content melts with a low oxidizing gas flow rate*

A low flow rate of the oxidizing gas of 5–7.5 ml min^{-1} was used in these experiments to minimize any effect of impinging jet momentum on the phenomena. In view of the low flow rate, the reaction gas was passed for several minutes before interrupting the power to the induction furnace. No gas was passed down the second tube. The carbon, manganese and silicon concentrations were 0.0010–0.0013, 0.40–0.41 and 0.13–0.14 wt%, respectively. Small additions of aluminium were made for each set of experiments and the resulting concentration of aluminium was successfully maintained at about 0.003 wt%. The initial melt had a very low sulphur concentration of 0.009 wt%. Additions of iron sulphide were made to give two additional melts of 0.056 and 0.21 wt% S.

Despite the low aluminium concentration, the expanding arcs could be seen clearly and alumina accumulated away from the inlet tube. Figure 4 is a close-up view of the area near to the gas inlet tube, taken at a frame rate of 250 s^{-1}, from one of the experiments with the 0.2 wt% S melt and CO_2 as the oxidizing gas. The highlights on the expanding arc are consistent with the oblique illumination from the tip of the inlet tube being reflected by the moving surface disturbance.

Figure 5 shows, as the open circles, the pulsation frequency as a function of the CO_2/CO ratio in the oxidizing gas stream for the 0.009 wt% S melt. The two melts with higher sulphur concentrations at $p_{CO_2}/p_{CO} = \frac{1}{5}$ gave essentially the same value, as indicated by the open square. Within the experimental error, estimated to be about 10%, there appears to be no dependence on the sulphur concentration. The pulsation frequency for a pure CO_2 gas stream was greater than 2 s^{-1} for all of the sulphur concentrations, but the exact values could not be established with confidence.

Figure 6 shows the velocities of the 'necklaces' at approximately 10 mm from the

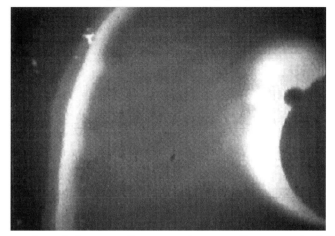

Figure 4. Close up view of the expanding arc, taken at a frame speed of $250\,\mathrm{s}^{-1}$, from an experiment with the 0.21 wt% S melt, showing what appears to be the surface disturbance, made visible by the oblique illumination from the tip of the gas inlet tube.

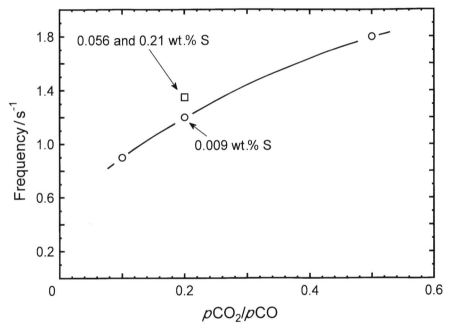

Figure 5. The pulsation frequency as a function of the CO_2/CO ratio in the gas stream for the 0.003 wt% Al melt at low gas low rates.

edge of the gas delivery tube as a function of the sulphur concentration for CO_2 and $p_{CO_2}/p_{CO} = \frac{1}{5}$ in the oxidizing gas stream. The experimental uncertainty in the velocity is estimated to be about 5%. There appears to be a distinct separation in the data and a clear reduction in the velocity with increasing sulphur concentration. The velocity observed in the experiments with high aluminium-content melts and high flow rates of the gas mixture with $p_{CO_2}/p_{CO} = \frac{1}{5}$ is shown by the closed circle in the same figure. There appears to be little difference in the velocity for the two conditions.

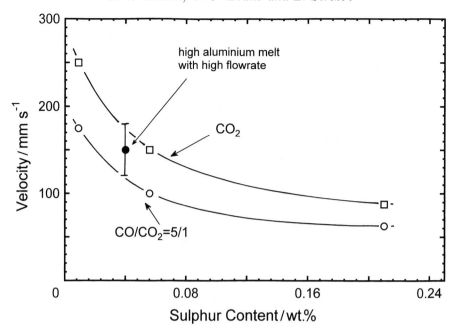

Figure 6. The velocities of movement at approximately 10 mm from the edge of the gas inlet tube in experiments with the 0.003 wt% Al melt and low gas flow rates.

In addition to the effect of sulphur on velocity, there was a marked tendency for the alumina oxidation product to form larger agglomerates at higher sulphur concentration. This can be seen by comparing figure 2 with figure 7, which is a sequence of frames, separated by 40 and 32 ms, in the order A → C, for experiments with CO_2 as the oxidizing gas and 0.21 wt% S in the steel. The larger agglomerates appear to be up to about 1 mm in diameter. Again, a slow drift of the surface from left to right was occurring during this particular sequence.

In the experiments with the low aluminium and highest sulphur content melts, irregular streaks appeared to follow the expanding arc. These can be clearly seen in figure 7c. From repeated observations, it is concluded that they most likely represented a second liquid phase which rapidly dissolved back into the bulk melt. Attempts to resolve this by the use of higher magnification (×7) were unsuccessful.

An addition of 0.04 wt% Se was made to the 0.21 wt% S melt to study the effect of a further decrease in the surface tension of the melt. Observations were made with a CO_2 reaction gas. Although pulsations of the surface near to the gas inlet tube occurred at an erratic frequency of about 1 Hz, no expanding 'necklaces' could be seen. Instead, a 'raft' of apparently solid oxidation product, approximately 5 mm in diameter, drifted from beneath the gas inlet tube at irregular intervals of 2–3 s. A detached 'raft' is shown in figure 8, where the tip of the gas inlet tube (8 mm OD) is at the right-hand side of the figure.

4. Discussion

(a) Origin of the phenomena

If the simplified case of liquid iron containing only aluminium as a deoxidant is considered, the reaction between the oxidizing gas stream and aluminium in a

(a)

(b)

(c)

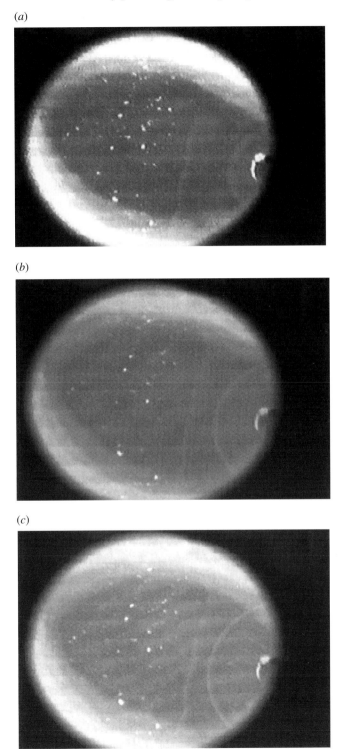

Figure 7. Sequence of frames (a)–(c), separated by 40 and 32 ms, in the order (a)–(c), from experiments with the 0.003 wt% Al melt with 0.21 wt% S at a low flow rate of CO_2 as the oxidizing gas.

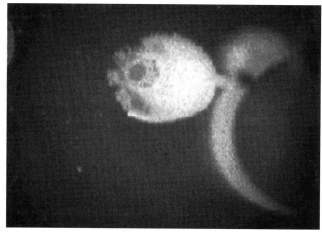

Figure 8. Detachment of a 'raft' of oxidation product from under the gas inlet tube during experiments with the 0.005 wt% Al melt containing 0.21 wt% S and 0.04 wt% Se with a low flow rate of CO_2. The frame speed is $500\ \text{s}^{-1}$.

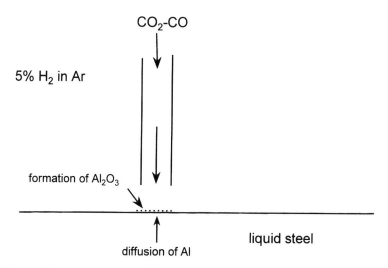

Figure 9. Schematic representation of the reaction between aluminium and the oxidizing gas stream, as assumed in the analysis.

stagnant melt may be represented schematically, as shown in figure 9. If there is no barrier to nucleation, alumina will be formed at the surface by the oxidation of aluminium diffusing from the bulk of the melt. For a constant flux of oxygen-containing species in the gas stream, and considering only one-dimensional diffusion, valid near to the centre line, the concentration of aluminium at the surface, C_S, will be given by

$$C_S = C_B - \frac{2Ft^{1/2}}{(D\pi)^{1/2}}\left(\frac{2}{3}\right), \tag{4.1}$$

where C_B is the concentration of aluminium in the bulk, F is the flux of oxygen atoms from the reacting gas species, D is the chemical diffusivity of aluminium and t is the elapsed time. The stoichiometric requirement is met by the factor $\left(\frac{2}{3}\right)$.

The aluminium concentration as a function of reduced time will follow the curve

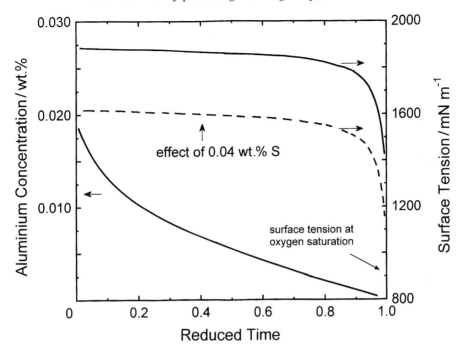

Figure 10. Calculated values of the aluminium concentration and surface tension below the centre of the gas inlet tube as a function of reduced time, from equations (4.1)–(4.3).

shown in figure 10. Turkdogan (1980) has shown that at concentrations of aluminium greater than 1 ppm, the deoxidation solubility product at $1600\,^{\circ}$C is given by

$$(a_{\mathrm{Al}})^2(a_{\mathrm{O}})^3 = 4.3 \times 10^{-14}, \tag{4.2}$$

where a_{Al} and a_{O} are the activities of aluminium and oxygen with respect to standard states of the 1 wt% ideal solutions. The surface tension of liquid iron as a function of the activity of oxygen at $1600\,^{\circ}$C, based on the work of Jimbo & Cramb (1992), is given by

$$\sigma = 1890 - 299\ln(1 + 100a_{\mathrm{O}}) \quad \mathrm{mN\,m}^{-1}. \tag{4.3}$$

Combination of equations (4.1)–(4.3) allows the calculation of the surface tension as a function of time. This is shown as a function of the reduced time in figure 10. Thus, following a small decrease in the surface tension for the bulk of the time, there is a sharp increase in the driving force for the radial flow of the liquid steel from under the impinging jet.

The local flux, F, of oxygen from the gas stream may be estimated from the modified Rao & Trass correlation due to Belton & Belton (1964) for impinging flow from a nozzle, namely,

$$Sh = 0.03Re^{1.06}\,Sc^{0.33}(x/d)^{-0.09}, \tag{4.4}$$

where the Sherwood number Sh is md/D, the Reynolds number Re is $du\rho/\mu$, the Schmidt number Sc is $\mu/\rho D$, x is the nozzle to surface distance, d is the inside diameter of the nozzle (tube), μ is the viscosity of the gas mixture, ρ is the density of the gas mixture, D is the interdiffusivity, u is the average velocity of the gas mixture and m is the mass transfer coefficient.

Taking the values of the transport properties of the gases from Turkdogan (1980), the derived value for the mass transfer coefficient of CO_2 is about 2.1 cm s^{-1} for the flow rate and gas composition of the first series of experiments. If $p_{CO_2} \to 0$ at the interface,

$$F = 1.4 \times 10^{-6} \text{ mol cm}^{-2} \text{ s}^{-1}. \tag{4.5}$$

Substituting this value in equation (4.1), and taking the diffusivity of aluminium in liquid iron to be 5×10^{-5} cm^2 s^{-1} from the recent work of Kawakami *et al.* (1997), leads to a value for t of 0.11 s for the condition $C_S \ll C_B$. Thus, following the radial dispersal of the oxidation product and low surface tension liquid, and the concomitant upward flow of fresh bulk liquid to the surface, the process would be expected to be successively repeated at a frequency of about 9 Hz.

Despite the approximations in this treatment, the order of magnitude agreement with the observed frequency of 4–5 Hz strongly suggests that the mechanism is essentially as described.

In view of the approximate additivity of the effects of oxygen and sulphur on the surface tension at low sulphur concentration (Ogino *et al.* 1983), discussed later, the effect of about 0.04 wt% S should be simply to lower the change in surface tension to that indicated by the dashed curve in figure 10. Similarly, the elapsed time is not strongly dependent on the 'end point'. If, for example, the dispersal of the alumina product occurs when the surface tension is depressed by 100 mN m^{-1} rather than by, say, 500 mN m^{-1}, the shortening in the elapsed time would only be by about 10%.

The second series of experiments used very low flow rates of oxidizing gas, well below the range of applicability of the correlation (equation (4.4)). Linear gas velocities were in the range of only 0.5–1.0 cm s^{-1}. Accordingly, it is not possible to make an experimentally well-established estimate of the flux of oxygen to the surface of the melt. However, the oxidation appears to be confined to the region directly under the gas inlet tube, as indicated by the size of the 'raft' of oxidation product in figure 8.

If, to a very first approximation, the average flux of oxygen to the surface, F, is taken as proportional to the volume flow rate of the gas, the ratio of the flux in the second series of experiments to that in the first series of experiments at $p_{CO_2}/p_{CO} = \frac{1}{5}$ would be about 0.07. The bulk concentration of aluminium, C_B, is about 0.003 wt%. According to equation (4.1), the pulsation frequency, $1/t$, should therefore be about 2 Hz, i.e. reduced by a factor of about 5. The observed frequency for these conditions (figure 5) was about 1.2 Hz. Thus, the lowering of the pulsation frequency is reasonably consistent with the proposed mechanism.

Within the experimental error, there was no effect of sulphur concentration on the pulsation frequency at $p_{CO_2}/p_{CO} = \frac{1}{5}$, as shown in figure 5. Belton (1993) has shown that sulphur markedly decreases the interfacial rate of dissociation of CO_2 on liquid iron, this observation is consistent with the transport of oxygen to the surface being dominated by physical mass transfer processes, as assumed in analysis.

(b) The velocity of motion and the effects of changes in bulk surface tension

The effect of the concentration of sulphur on the surface tension of liquid iron at 1550–1600 °C, based upon the bulk of the available experimental data, has been shown by Belton (1976) to be described by the equation

$$\sigma = 1778 = 195 \ln(1 + 185 a_S) \quad \text{mN m}^{-1}, \tag{4.6}$$

where a_S is the activity of sulphur with a standard state of the 1 wt% ideal solution. Neglecting the relatively small effects of the other solutes on the activity coefficient

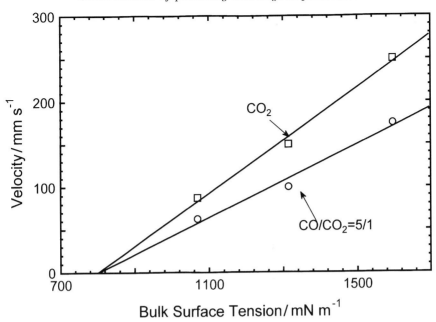

Figure 11. Velocities of the front movement at 10 mm from the edge of the gas inlet tube as a function of the calculated bulk surface tension for the experiments with the 0.003 wt% Al melt.

of sulphur, the results in figure 6 may therefore be converted to a plot of the observed velocity versus the surface tension of the bulk liquid iron, as shown in figure 11. The results for both oxidizing conditions appear to extrapolate to zero velocity at about 800 mN m^{-1}.

Naidich & Zabuga (1992) have derived the following expression for the Marangoni-driven velocity of front movement when a droplet, enriched with a surface active solute, is introduced to the surface of a liquid:

$$v = (\Delta\sigma^2/4\alpha^2\mu\rho r)^{1/3}, \tag{4.7}$$

where α is a constant equal to 0.332, $\Delta\sigma$ is the difference in surface tension between the bulk liquid and the droplet, μ and ρ are the viscosity and density of the liquid, and r is the distance between the moving front and the source. Reasonably good agreement was found between observed and calculated velocities for experiments with water and surfactant-enriched droplets.

Hirashima *et al.* (1995) have recently used computation fluid dynamics to model the continuous Marangoni-induced flow caused by the dissolution of nitrogen into liquid iron from an impinging jet. Agreement between their measured and calculated velocities was obtained when an effective viscosity, defined by

$$\mu_{\text{eff}} = \mu + \mu_{\text{t}}, \tag{4.8}$$

was used in the computation, where μ_{t} is the turbulent viscosity, taken to be about 5μ. This was considered to be reasonable for mildly turbulent flow.

Taking the value of μ to be 5×10^{-3} Pa s, and assuming a similar multiplicative factor for the turbulent viscosity, substitution in equation (4.7) leads to

$$v \approx 1.03(\Delta\sigma)^{2/3} \quad \text{m s}^{-1} \tag{4.9}$$

at a distance of 10 mm from the source, when the units for $\Delta\sigma$ are N m^{-1}.

The trend of decreasing velocity with decreasing bulk surface tension in figure 11 is therefore to be expected if $\Delta\sigma$ similarly decreases. There is experimental evidence that this is true for a given change in oxygen activity. Gaye *et al.* (1984) have shown that at a sulphur concentration of about 0.1 wt% or above, the further depression of the surface tension caused by increasing the oxygen activity from *ca.* 0 to 0.004 is hardly detectable. Whereas, at very low sulphur concentration (less than 0.03 wt%), the effects of oxygen and sulphur have been found to be approximately additive (Ogino *et al.* 1983).

The extrapolation to zero velocity at a bulk surface tension of about $800 \ \mathrm{mN \ m^{-1}}$ suggests that this is the region where an increase in the activity of oxygen has no significant effect on the surface tension. Unfortunately, there have been no independent studies of the effect of high activities of oxygen on the surface tension of liquid iron at high sulphur concentrations. Kozakevitch & Urbain (1961) reported that the addition of 0.04 wt% of selenium to liquid iron reduces the surface tension to about $1000 \ \mathrm{mN \ m^{-1}}$. The quantitative effect of the addition of this amount to the 0.2 wt% S melt has not been experimentally determined, but it is reasonable to expect that the resulting surface tension would be significantly below $1000 \ \mathrm{mN \ m^{-1}}$. The observation that the 'raft' of oxidation product did not disperse under these conditions is consistent with the driving force, $\Delta\sigma$, being very small.

If it is assumed that the dispersal of the alumina occurs just before the nucleation of a liquid manganese–silicate oxidation product, the maximum activity of oxygen would be about 0.015 (Turkdogan 1980). At low sulphur concentration, the depression of the surface tension can be estimated from equation (4.3) to be about $270 \ \mathrm{mN \ m^{-1}}$. Substitution of this in equation (4.9) gives an expected velocity of movement of the front of about $0.43 \ \mathrm{m \ s^{-1}}$ at a distance of 10 mm from the source. This is to be compared with the observed values for the 0.009 wt% S melt of about $0.25 \ \mathrm{m \ s^{-1}}$ for CO_2 as the oxidizing gas, and $0.16 \ \mathrm{m \ s^{-1}}$ for $p_{CO_2}/p_{CO} = \frac{1}{5}$. Thus the magnitude of the velocity is reasonably consistent with the assumptions, if equation (4.7) is accepted.

The observed increase in the velocity by a factor of about 1.5 for a six-fold increase in the partial pressure of CO_2 in the gas stream suggests that there is some sensitivity of the activity of oxygen, and hence $\Delta\sigma$, at the point of dispersal of the oxidation product on the flux of oxygen to the surface. Neglecting the uncertainties in the measurements, a change in the end-point activity of oxygen by a factor of about two would account for the difference for the low sulphur melt, if the value of $\Delta\sigma$ is of the order of $102 \ \mathrm{mN \ m^{-1}}$. However, the similarity of the velocity at a given concentration of sulphur between the first and second series of experiments is consistent with the bulk surface tension being the most important factor in establishing the velocity under the conditions of the experiments.

5. Conclusion

Consistent with the observations, the foregoing analysis has contained the assumption that the 'raft' of alumina oxidation product remains intact under the gas delivery tube until the Marangoni driving force is sufficient to cause the sudden dispersion. It has been shown recently by Emi *et al.* (1997) and Yin *et al.* (1997) that relatively long-range ($\leqslant 50 \ \mu\mathrm{m}$) forces exist, probably due to capillary effects, between alumina particles on the surface of liquid iron. These cause the particles to agglomerate.

It appears likely that these are the forces that prevent the random dispersal of the alumina until possibly a critical value of $\Delta\sigma$ is exceeded.

References

Belton, G. R. 1976 *Metall. Trans.* B **7**, 35–42.

Belton, G. R. 1993 *Metall. Trans.* B **24**, 241–258.

Belton, G. R. & Belton, R. A. 1980 *Trans. Iron Steel Inst. Jap.* **20**, 87–91.

Emi, T., Yin, H. & Shibata, H. 1997 *CAMP-ISIJ* **10**, 93–96.

Gaye, H., Lucas, L.-D., Olette, M. & Riboud, P. V. 1984 *Can. Met. Q.* **23**, 179–191.

Hammerschmid, P. 1987 *Stahl Eisen* **107**, 35–40.

Hirashima, N., Choo, R. T. C., Toguri, J. M. & Mukai, K. 1995 *Metall. Mater. Trans.* B **26**, 971–980.

Jimbo, I. & Cramb, A. W. 1992 *ISIJ Int.* **32**, 26–35.

Kawakami, M., Yokoyama, S., Takagi, K., Nishimura, M. & Kim, J.-S. 1997 *ISIJ Int.* **37**, 425–431.

Kozakevitch, P. & Urbain, G. 1961 *Mem. Sci. Rev. Met.* **58**, 517–534.

Mukai, K. 1992 *ISIJ Int.* **42**, 19–25.

Naidich, Yu. V. & Zabuga, V. V. 1992 *Metally* no. 4, 40–46.

Ogino, K., Nogi, K. & Hosoi, C. 1983 *Tetsu-to-Hagane* **69**, 1988–1994.

Olette, M. 1993 *ISIJ Int.* **33**, 1113–1124.

Rao, V. V. & Trass, O. 1964 *Can. J. Chem. Engng* **42**, 95–99.

Sigworth, G. K. & Elliott, J. F. 1974 *Met. Sci.* **8**, 298–310.

Turkdogan, E. T. 1980 *Physical chemistry of high temperature technology.* New York: Academic.

Yin, H., Shibata, H., Emi, T. & Suzuki, M. 1997 *ISIJ Int.* **37**, 936–945.

Marangoni enhancement of desulphurization of liquid iron

By D. R. Gaskell and A. Saelim

*School of Materials Engineering, Purdue University,
West Lafayette, IN 47907, USA*

Liquid iron at 1600 °C, containing 0.088 wt% S and 0.070 wt% O, was desulphurized by a lime-saturated calcium ferrite slag at a rate significantly greater than that calculated on the assumption that diffusion of sulphur in the liquid iron was rate controlling. It is postulated that the transfer of sulphur across the slag–metal interface caused local variations in the interfacial energy, which induced Marangoni stirring in both the metal and slag phases.

Keywords: desulphurization; liquid iron; lime-saturated calcium ferrite slag; kinetics; Marangoni stirring; turbulence

1. Introduction

When placed in a slip-cast lime crucible at 1600 °C, liquid irons containing 1.62 and 0.66 wt% sulphur, were desulphurized by the crucible. Reaction between the crucible and sulphur in the iron produces a liquid which rapidly penetrates the wall of the crucible. The assumption that the rate of desulphurization of the melt was controlled by diffusion of sulphur in the iron gave a diffusion coefficient of $6.7 \pm 1.7 \times 10^{-5}$ cm^2 s^{-1}, which is in good agreement with the values of 4.86×10^{-5} and 4.63×10^{-5} cm^2 s^{-1} measured at 1600 °C, respectively, by Kawai (1956) and Majdic *et al.* (1969) using conventional experimental techniques. It was found that liquid iron containing 0.088 wt% sulphur was not desulphurized by the crucible, presumably because the activity of iron sulphide in this melt was not high enough to cause reaction with the lime crucible. However, when brought into contact with lime-saturated liquid iron oxide (of composition 58% FeO–42% CaO) at 1600 °C, the melt was desulphurized at a rate significantly higher than that calculated on the assumption of control by diffusion of sulphur in the liquid iron. For example, with a fixed geometry of a slag–metal interfacial area of 1.5 cm^2 and a mass of iron melt of 5 g, the measured sulphur content of the iron decreased from 0.08 to 0.02 wt% in 4 min in comparison with a decrease from 0.08 to 0.065 wt% in 4 min calculated assuming diffusion control in the liquid iron. The addition of 4 wt% CaF$_2$ to the lime-saturated oxide melt had no influence on the rate of desulphurization of the iron. It is postulated that this enhanced rate of desulphurization is caused by Marangoni turbulence at the slag–metal interface and the following is an application, to the experimental system, of the two fundamentally differing forms of the Marangoni effect given by Berg (1982). The transfer of sulphur from metal to slag causes a decrease in the interfacial tension and, in the first type of effect, any local variation in the interfacial tension causes a disturbance in the form of a dilation, as illustrated in figure 1a. This dilation, in turn, brings solute-rich metal to the site of the dilation which further decreases the surface tension and sets

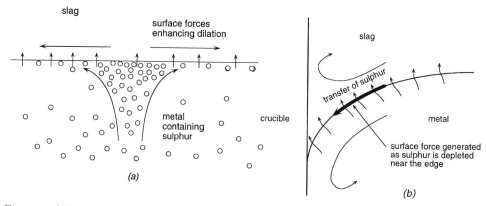

Figure 1. Schematic representations of (*a*) the first form of the Marangoni effect and (*b*) the second form of the Marangoni effect (adapted from Berg (1982)).

Figure 2. Illustration of the growth of a 'head' on the iron ingot with time caused by dissolution of the CaO crucible along its line of contact with the slag–metal interface. Desulphurization times increase in passing from the ingot on the left to the ingot on the right.

up forces which promote further dilation until macroscopic convection is generated. In the second type of effect, the transfer of sulphur upward from the interface preferentially depletes a narrow region beneath the meniscus near the crucible wall, which causes a higher interfacial tension near the wall. This propels the interface toward the wall and sets up the circulation in the bulk phase shown in figure 1*b*. Evidence for the latter effect was provided by the observation of the growth of a 'head' on the metal ingot during desulphurization. This growth is shown in figure 2, in which the reaction time increases from left to right. The enrichment, in iron and sulphur, of the slag being circulated at the wall undersaturates the slag with respect to CaO and hence causes dissolution of the crucible along its line of contact with the slag and metal.

References

Berg, J. C. 1982 *Can. Met. Q.* **21**, 121.
Kawai, Y. 1956 *J. Jap. Inst. Metals* **20**, 514.
Majdic, A., Graf, D. & Schenck, H. 1969 *Arch. Eisenhuttenw.* **40**, 627.

The capillary effect promoting collision and agglomeration of inclusion particles at the inert gas–steel interface

By H. Shibata, H. Yin and T. Emi

*Institute for Advanced Materials Processing, Tohoku University,
Sendai 980-77, Japan*

Gas bubbling and flux injection processes have been extensively used in metallurgical industries for several decades since the processes have been confirmed to be effective in removing impurity elements and non-metallic inclusions from liquid metals and improving the morphology of inclusions. Aiming at more effective use of gas and flux powder for injection refining to obtain higher quality products at lower cost, elucidation of the interfacial phenomena at inert gas–solid particle–liquid metal boundaries has become increasingly important. However, difficulties inherent to high-temperature experiments have prevented progress in this field. A new attempt is made in this paper to reveal the collision and agglomeration behaviour of solid and liquid inclusions and flux powders at gas–steel melt interface.

Keywords: capillary interaction; inclusion; particle; steel melt; interface;
agglomeration; laser microscope

1. Introduction

In our previous work, a confocal scanning laser microscope (CSLM) combined with an infrared image furnace (Chikama *et al.* 1996) was used as a new investigating tool, which has made it possible to carry out an *in situ* real-time observation of collision and agglomeration behaviour of inclusion particles on the surface of steel melts at high temperatures. On molten steel surfaces, a long-range strong attraction, which can extend to tens of micrometres, was clearly observed to operate between solid inclusion particles (Yin *et al.* 1997). The strong attraction results in a quick agglomeration of solid particles at the inert gas–steel melt interface, and hence enhances the growth and removal of solid inclusions as well as the reactions between flux particles and inclusions on bubble surface in injection metallurgy practice. It was also found that there was no such strong attraction between liquid inclusion particles on molten steel surfaces (Yin *et al.* 1997). The present work is intended to make clear the origin of the attraction force between inclusions partly immersed in the steel melt surface.

2. Experiment

A mirror-polished disc specimen (4.3 mm diameter × 2 mm) of steel was set into a high purity alumina crucible and placed in a 650 W image furnace chamber (described in detail in Chikama *et al.* (1996)). Before experiments, the furnace chamber, specimen and crucible were evacuated to 10^{-5} Torr for desorbing gases. Then, the specimen and crucible were baked at about 500 K for 1 h to remove absorbed moisture

143

Table 1. *Compositions of steels studied and inclusion particles surfaced (mass%)*

grade name	concentration of elements in steel							composition of inclusions			note
	C	Si	Mn	P	S	sol Al	total O	CaO	Al_2O_3	SiO_2	
LCAK	0.038	0.01	0.20	0.010	0.010	0.035	0.0200	—	100	—	solid
Fe–3%Si	0.0006	3.35	0.01	0.001	0.001	0.001	0.0024	—	> 80	< 20	solid
HC–Ca	0.840	0.17	0.50	0.010	0.010	0.001	0.0030	< 20	> 80	—	solid
SK	0.001	0.45	0.13	0.002	0.003	0.001	0.0103	< 5	< 2	> 95	solid
HC–Ca	0.840	0.17	0.50	0.010	0.010	0.001	0.0030	~ 50	~ 50	—	liquid
SK	0.001	0.45	0.13	0.002	0.003	0.001	0.0103	~ 25	~ 50	~ 25	liquid
HSLA	0.048	0.31	1.17	0.006	0.0005	0.036	0.0015	> 30	~ 60	< 10	liquid

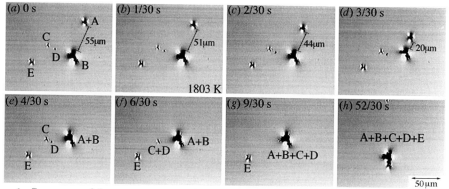

Figure 1. Sequence of long-range attraction, quick agglomeration and the cluster formation process of alumina inclusion particles observed on a LCAK steel melt surface at 1803 K with CSLM.

under the flow of highly purified Ar gas containing O_2, H_2O, CO_2 and CO gases of less than 10 ppb. After that, the specimen was heated to its melting point (usually *ca.* 1800 K) at 100 K min^{-1}. The experiment was carried out under the purified Ar atmosphere which was further deoxidized by pure titanium, heated to *ca.* 1700 K to prevent oxidation of the specimen surface. During the experiment, the dynamic pictures of the specimen surface, scanned by a He-Ne laser beam and monitored by CCD image sensor, were displayed on a CRT screen and recorded on a video-tape at $\frac{1}{30}$ s intervals (Yin *et al.* 1997). Interesting behaviour of inclusion particles is revealed and interparticle attraction forces deduced by analysing the recorded video image frame by frame. Composition of inclusion particles on the melt surface was determined by EPMA after quenching the molten specimen. The compositions of steel grades studied and the inclusions found on the specimen surfaces are listed in table 1.

3. Results

After the steel specimen was melted, many small inclusion particles surfaced from the steel bath. They quickly agglomerated and formed large clusters on the bath surface due to a long-range attraction operating between them. Various solid inclusion

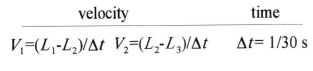

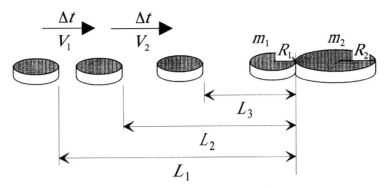

Figure 2. Schematic diagram to calculate acceleration from observed change in the position of disc-like inclusion particles.

particles, such as Al_2O_3, Al_2O_3–80 mass%SiO_2, CaO–80%Al_2O_3 and CaO–Al_2O_3–95%SiO_2, etc., observed on various steel melt surfaces, attracted each other strongly over a distance of a few tens of μm (described in detail in Yin *et al.* (1997)). An example of such an accelerated attraction between pure alumina particles A and B on LCAK molten steel surface is shown in figures 1*a–e*. The distance between the two particles is quickly decreased by the attraction from 55, 51, 44 to 20 μm at $\frac{1}{30}$ s intervals. The formation of clusters as induced by this attraction from A + B, C + D, (A + B) + (C + D) to A + B + C + D + E is shown in figures 1*e–h*.

To elucidate the nature of the long-range attraction, an approximate calculation of the magnitude of the attraction force is carried out using the observed acceleration with the estimated mass of the particles. The particle shape is approximated to be an elliptical disc, and the length of the long axis (d_1) and the short axis (d_2) of a particle is noted. Also, the height of all the discs is taken to be 2 μm on the basis of confocal observation of the thickness of alumina particles on the steel melt surface. Further, the elliptical disc is approximated to be circular, of radius R, which is taken to be equal to the geometric average, $R = \frac{1}{2}\sqrt{d_1 d_2}$. The acceleration, a_1, of the guest particle when the host particle in the particle pair stayed quiescent is determined from the change of the position of the guest particle at $\frac{1}{30}$ s intervals, as shown in figure 2. The attraction force, F, is then given by

$$F = m_1 a_1,\tag{3.1}$$

where m_1 is the mass of the guest particle. In this force analysis, the frictional force which arises from viscous drag of the particle by the liquid steel surface is ignored. Thus, the apparent attraction force operating between alumina solid particles is determined to be in the range of 10^{-16}–10^{-14} N, as shown in figure 3. Also, the acting length of the attraction, defined as L_1 in figure 2, is observed to be longer than 10 μm for particles with radius of larger than 2 μm. The force and its acting distance increase with the size of particle in a particle–particle pair, as shown in

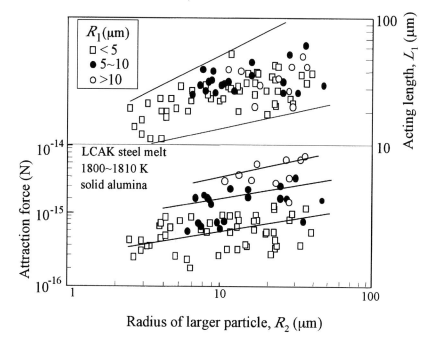

Figure 3. Observed long-range attraction force and its acting length versus size of alumina particle.

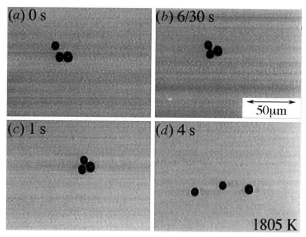

Figure 4. Touching and separation of liquid CaO–50%Al$_2$O$_3$ particles with random surface flow on an HC–Ca steel sample.

figure 3. In addition, the attraction force and acting length are not significantly influenced by surfactant elements, like sulphur, in molten steel (Yin *et al.* 1997).

On the other hand, no long-range attraction was observed on molten steel surfaces between all the kinds of liquid inclusion particles (Yin *et al.* 1997) listed in table 1. A typical case is shown in figure 4 where three 50%CaO–50%Al$_2$O$_3$ particles on the surface of HC–Ca steel melt come very close to each other as they are driven by a random surface flow. However, they move away subsequently without showing any attraction in between. According to our preliminary investigation (Yin *et al.* 1997), such long-range strong attraction cannot be attributed to the van der Waals force,

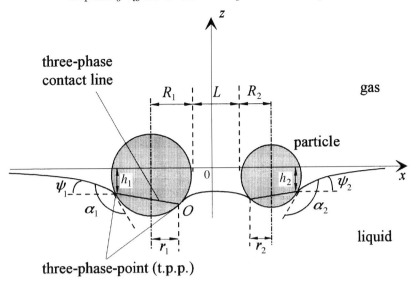

Figure 5. Schematic diagram of capillary meniscus around two spherical particles.

the Rayleigh–Benard flow of the steel melt (Pearson 1958), macroscopic local surface flow on the steel melt (Yin *et al.* 1997), the laser trapping effect (Ashkin 1970) or to coulombic interaction between solid inclusions on the steel melt surface.

4. Discussion

Capillarity causes a strong attraction between two floating solid bodies partly immersed in water when they come very close to each other. In terms of surface physics, the observed attraction between the solid inclusion particles partly immersed in molten steel may be the same as that in the gas–water system. An important parameter, capillary length, $q^{-1}(= [\Delta\rho g/\gamma]^{-1/2})$, was defined by Kralchevsky *et al.* (1994) to characterize the range of capillary interaction between solid particles at gas–liquid interfaces, and calculated to be 2.7 mm for the gas–water system. Here, $\Delta\rho = \rho_{\mathrm{L}} - \rho_{\mathrm{G}}$ (ρ_{L} and ρ_{G} are the density of liquid and gas phases), g is the acceleration due to gravity and γ is the surface tension of the liquid. According to this definition, the capillary length of an inert gas–molten steel system can be calculated to be 5.1 mm as the density and surface tension of molten steel are $7000 \ \mathrm{kg \ m^{-3}}$ and $1.8 \ \mathrm{N \ m^{-1}}$ (density of gas is negligible here). This comparison shows that the range of capillary interaction of a gas–steel melt system is longer than that of a gas–water system.

To confirm that the observed attraction is also caused by the capillary effect, the calculated order of magnitude of the capillary attraction must be the same as the observed order of magnitude. Paunov *et al.* (1993) theoretically derived an equation for the capillary interaction energy between two spherical particles 1 and 2 at a gas–liquid interface when their radius, R_i, is smaller then $100 \ \mu$m as

$$\Delta W = -\pi\gamma \sum_{i=1}^{2} (Q_i h_i - Q_{i\infty} h_{i\infty})[1 + O(q^2 R_i^2)], \quad i = 1, 2, \tag{4.1}$$

where subscript i represents the particles 1 and 2, ∞ means that the separation between the two particles is infinite and $O(y)$ is the zeroth-order function of approximation. According to a series of equations derived by Paunov *et al.* (1993), Q_i

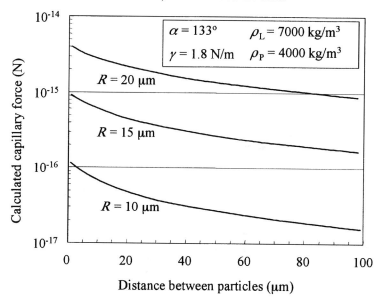

Figure 6. Dependence of calculated capillary force on the distance between two identical particles with varied size.

(defined as $r_i \sin \psi_i$; r is the radius of the three-phase contact line and ψ is the angle of meniscus slope) and the height difference (h_i shown in figure 5) at the separation, L, between the two particles can be calculated numerically.

Furthermore, when $L \to \infty$, $Q_{i\infty}$ was given by Chan *et al.* (1981) as

$$Q_{i\infty} = \tfrac{1}{6}q^2 R_i^3 (2 - 4D_i + 3\cos\alpha_i - \cos^3\alpha_i)[1 + O(qR_i)]. \qquad (4.2)$$

Also, $h_{i\infty}$ is given by (Paunov *et al.* 1993)

$$h_{i\infty} = r_{i\infty}\sin\psi_{i\infty} \ln \frac{4}{\gamma_e q r_{i\infty}(1 + \cos\psi_{i\infty})}, \quad (qr_{i\infty})^2 \ll 1, \qquad (4.3)$$

where D is the density ratio of particle to liquid, α is the contact angle of particles and $\ln\gamma_e$ is the Euler–Masceroni constant.

Once ΔW is calculated at different L, the capillary interaction force is given by

$$F = \mathrm{d}(\Delta W)/\mathrm{d}L. \qquad (4.4)$$

The value of F calculated from equations (4.1)–(4.4) with the data from a gas–alumina–molten steel system for a pair of two identical spherical particles of $R = 10$, 15 or 20 µm is shown in figure 6, where the capillary interaction between the particles is shown to be long range. This result agrees well with the present observation. For larger particles ($R > 20$ µm), the calculated force is of the same order of magnitude as that observed in figure 3, despite the difference in particle shape assumption between the observation and calculation. For smaller particles ($R < 10$ µm), however, the calculated force is one or two orders of magnitude smaller than the observed force in figure 3. If particle shape and viscous drag force are taken into account, the difference can be greater. Figure 7 compares observed with revised calculated data, where the particles are considered to be disc-like (Yin *et al.* 1998). Although the calculated data are two orders of magnitude smaller than the observed data, the fact that the slopes of the two lines are almost the same shows that the two forces share the same nature (Yin *et al.* 1998).

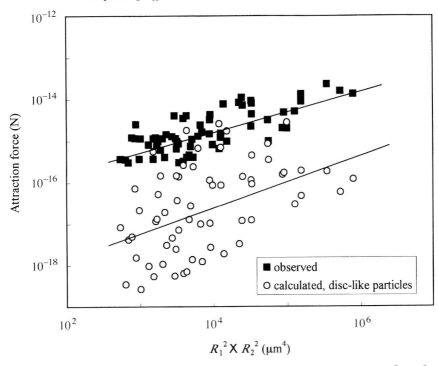

Figure 7. Comparison of calculated force with observed force versus $R_1^2 \times R_2^2$.

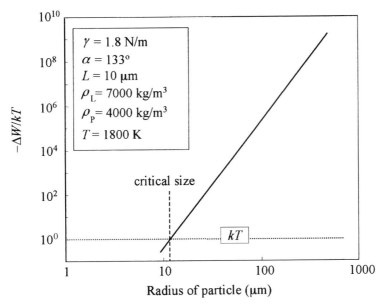

Figure 8. Chart of $-\Delta W/kT$ versus particle radius, R, for two identical particles at the gas–molten steel interface.

When the capillary interaction energy is compared with the thermal energy, kT (k is the Boltzmann constant and $T \approx 1800$ K), for a gas–alumina–molten steel system, the critical size of the particles, below which the capillary interaction is negligible,

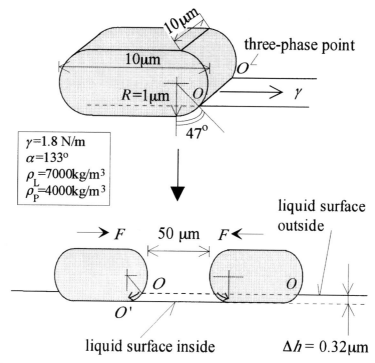

Figure 9. Schematic diagram of capillary attraction force calculation between particles on the melt surface by assuming that the liquid surface in between depresses down to the particle bottom.

can be estimated from figure 8. It is obvious here that, for particles with radii smaller than 15 μm, the capillary attraction is insignificant between alumina particles at the molten steel surfaces when the separation is greater than 10 μm. This conclusion is totally different from the present experimental observation. Actually, in most cases, alumina particles with radii of a few μm, like particles C and D in figure 1, are attracted to each other from about 10 μm away.

Some experimental data are available to support the above theoretical analysis for a system of two vertical cylinders of submillimetre diameter partly immersed in water (Velev 1993), which is fundamentally not that different from the two-sphere system. The theoretical calculation is in good agreement with the observation only at large separations (> 500 μm) between the two cylinders (Velev *et al.* 1993). For smaller separations, the observed capillary force is always stronger than the calculated capillary force because the linear superposition approximations made in the theoretical analysis cannot describe the capillary force well. The reason is that the superposition approximations are no longer linear when the separation gets small (Velev *et al.* 1993). However, no data have been reported to date to confirm the validity of the theory for the system of very small particles ($R < 20$ μm) at small separations of less than 50 μm.

Under these circumstances, a bold assumption is made for the system of small particles at small separations in the following force analysis. From the Laplace capillary equation (Bikerman 1970) and force balance analysis on particles (Schulze 1977) at steel melt surfaces, the height, h in figure 5, is calculated to be small (3.8 nm) for a spherical alumina particle with a radius of 10 μm. This result shows that the change of melt surface profile can be totally ignored when a small alumina particle (of the order of a few μm) is located at a molten steel surface. Thus, a more precise example is depicted in figure 9, where a $10 \times 10 \times 2 \, \mu m^3$ cuboid alumina particle with round edges is partly immersed in the steel melt surface.

When another cuboidal alumina particle comes close, at a distance of 50 μm, the steel melt surface in between will be dragged downward as the result of the capillary effect, leading to the depression of the three-phase point (TPP) O in figure 9. As the capillary depression is so large, a crude assumption is made here that the TPP will shift from O to the bottom, O', of particles in figure 9. The capillary force can then be calculated by (Bikerman 1970)

$$F = 0.5g(\rho_L - \rho_P)w\Delta h^2, \tag{4.5}$$

where, w is the width of the cuboid particle. After the same force balance analysis as mentioned above, the height is calculated to be 0.32 μm. In this case, the capillary attraction force, F, is calculated from equation (4.1) to be 1.5×10^{-14} N by ignoring the three-dimensional effect of the liquid surface. When the drag force of the melt surface acting on the particles during movement is subtracted, the capillary attraction becomes of the same order of magnitude as the observed attraction. The absence of long-range attraction between lens-like liquid particles at molten steel surfaces can be understood because the depression of the molten steel surface between the liquid particles cannot take place in view of their contact angles with molten steel, even when they come very close to each other.

Reservation should be made, however, that the shift of point O is dynamic when two particles come closer. Then, determining the precise position of the TPP in the dynamic process becomes complicated. Also, at the beginning of particle movement, the balance of capillary attraction and viscous drag force is difficult to analyse. Therefore, the estimation of the acting length of the capillary attraction remains an approximation.

5. Conclusions

In situ observed long-range attraction between solid particles on molten steel surfaces is mainly attributed to the capillary interaction when they get close to each other. This effect is useful in high-temperature metallurgy processes because it can accelerate the agglomeration of solid inclusion particles and reaction between solid inclusion particles and flux particles at the gas–molten metal interface.

References

Ashkin, A. 1970 *Phys. Rev. Lett.* **24**, 156.

Bikerman, J. J. 1970 *Physical surfaces*, ch. 1. New York: Academic.

Chan, D. Y. C., Henry, J. D. & White, L. R. 1981 *J. Colloid Interf. Sci.* **79**, 410.

Chikama, H., Shibata, H., Emi, T. & Suzuki, M. 1996 *Mat. Trans. JIM* **37**, 620.

Kralchevsky, P. A., Denkov, N. D., Paunov, V. N., Velov, O. D., Ivanov, I. B., Yoshimura, H. & Nagayama, K. 1994 *J. Phys. Condens. Matter* **6**, 395.

Paunov, V. N., Kralchevsky, P. A., Denkov, N. D. & Nagayama, K. 1993 *J. Colloid Interf. Sci.* **157**, 100.

Pearson, J. R. A. 1958 *J. Fluid Mech.* **4**, 489.

Schulze, H. J. 1977 *Int. J. Mineral Proces.* **4**, 241.

Velev, O. D., Denkov, N. D., Paunov, V. N., Kralchevsky, P. A. & Nagayama, K. 1993 *Langmuir ACS* **9**, 3702.

Yin, H., Shibata, H., Emi, T. & Suzuki, M. 1997 *ISIJ Int.* **37**, 936 and 946.

Yin, H., Emi, T., Shibata, H. & Kim, J. S. 1998 *2nd Int. Conf. on High Temperature Capillarity, Cracow, Poland, 29/6–2/7/1997.* (In the press.)

Discussion

J. C. EARNSHAW (*Department of Pure and Applied Physics, University of Belfast, UK*). I have a comment and a question regarding Professor Emi's experiments.

(1) Capillary forces will, indeed, provide long-ranged attractive forces between solid objects on a liquid surface (Gifford & Scriven 1971; Chan *et al.* 1981). Their range, typified by the capillary length $\sqrt{\sigma/g(\rho_s - \rho_l)}$ (where σ is the surface tension and $\rho_{s,l}$ are the densities of the solid and liquid), can be of the order of 1 cm. While gravity plays an essential role in the conventional model (Gifford & Scriven 1971), Lucassen has recently shown that for irregular particles there is an extra contribution which is independent of gravity (Lucassen 1992). This force can become repulsive at very short distances. This may be significant in Professor Emi's experiment, as it could provide the freedom to restructure demonstrated by the aggregates in his video.

(2) Some of Professor Emi's graphs appear to show an attractive force which *increases* with increasing separation. If this is indeed the case, how are we to understand such an unusual behaviour?

T. EMI. (1) The capillary length (here ρ_s and ρ_l in the equation gievn in the comment should be replaced by ρ_G and ρ_L) in our case is calculated and given (§4) to be 5.1 mm. For the formation of aggregates and the subsequent aggregate clustering, the acting range of capillary forces is much longer than the repulsive forces after Lucassen, as evidenced in my video presentation at the Royal Society.

The repulsive force may play a role in the early stage of cluster densification when the aggregates are about to stick together at their tips under quiescent circumstances, where the capillary attraction may compare with the repulsive force, if any. In the later densification stage, however, sintering between the stuck aggregates in the cluster plays a more important role.

(2) I regret that the graphs presented were misleading. In reality, larger particles in a particle pair exhibit stronger attraction at increased distances. We were aware of this difficulty in the graphs, and rearranged them in our paper by taking the particle size as a parameter.

Additional references

Gifford, W. A. & Scriven, L. E. 1971 *Chem. Engng Sci.* **26**, 287.

Lucassen, J. 1992 *Colloids Surf.* **65**, 131.

Pyrometallurgical significance of Marangoni flow: mechanism and contributions to processing

By T. Takasu and J. M. Toguri

Chemical Metallurgy Group, Department of Metallurgy and Materials Science,
University of Toronto, 184 College Street, Toronto,
Ontario, Canada M5S 3E4

In many pyrometallurgical refining processes, Marangoni convection plays an important role because it enhances the mass transport through the boundary layers. Marangoni flow is created by a surface-tension gradient by means of a compositional gradient or temperature gradient or an electrical potential gradient. A great deal of experimental difficulties exist at high temperatures to observe and to identify Marangoni flow from other competitive phenomena. However, this review discusses the mechanism and contribution of Marangoni flow to the following pyrometallurgical processes; the effect of changing the nitrogen composition on the surface of liquid iron, current efficiency in aluminium electrolysis, inclusion distribution in gas–metal arc welding, and the separation of metal or matte droplets from slag. It is concluded that an understanding of the role of Marangoni flow on the kinetics of these operations is important.

Keywords: metal refining; electrocapillary; aluminium electrolysis; welding

1. Introduction

Pyrometallurgical refining processes are concerned with the extraction of the valuable metals by the removal of impurities at high temperatures. The rate of the refining process invariably involves mass transfer through interfaces. The overall transfer rates are usually limited by mass transport through the boundary layers since the chemical reaction steps at high temperatures are sufficiently rapid.

A possible method to enhance mass transport in the interface region is by inducing Marangoni flow by creating a surface-tension gradient by means of a compositional gradient or temperature gradient or an electrical potential gradient.

A great deal of experimental difficulties exist at high temperatures to observe and to identify Marangoni flow from other competitive phenomena. However, this paper will discuss the mechanism and contribution of Marangoni flow to the following pyrometallurgical processes; the effect of changing the nitrogen composition on the surface of liquid iron, erosion rates of refractory, current efficiency in aluminium electrolysis, inclusion distribution in gas–metal arc welding and the separation of metal or matte droplets from slag.

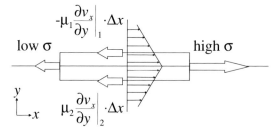

Figure 1. Schematics of the balance of the tangential stress on an interface and the induced flow.

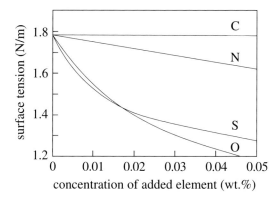

Figure 2. Influence of the addition of minor elements on the surface tension of liquid iron. These results were measured at 1823 K by Kozakevitch & Urbain (1961).

2. Fundamentals

Figure 1 shows the balance of the tangential stress on an interface and the induced flow. The difference in the interfacial tension induces a flow around the interface through the viscocity. The direction of the flow is from a region of low surface tension to a region of high surface tension.

Equation (2.1) is derived by taking a diffential form. The partial differential of the interfacial tension with respect to distance can be replaced by equation (2.2) since the interfacial tension changes with temperature, concentration and electric potential. The gradients of temperature, concentration and potential cause the flow to occur.

$$\frac{\partial \sigma}{\partial x} - \mu_2 \frac{\partial v_x}{\partial y}\bigg|_2 + \mu_1 \frac{\partial v_x}{\partial y}\bigg|_1 = 0, \tag{2.1}$$

$$\frac{\partial \sigma}{\partial x} = \frac{\partial \sigma}{\partial T}\frac{\partial T}{\partial x} + \frac{\partial \sigma}{\partial C}\frac{\partial C}{\partial x} + \frac{\partial \sigma}{\partial \phi}\frac{\partial \phi}{\partial x}. \tag{2.2}$$

In pyrometallurgical refining processes, the gradients of temperature and concentration are high because of the high operating temperatures and the existence of mass transport. An electric potential gradient occurs in an electrolysis cell due to the difference in current density. In addition, the application of an electric potential can be an alternative way to control the mass transfer rates when it is not possible to set up the gradients of temperature and/or concentration.

3. Examples in pyrometallurgical processes

(a) Gas–metal refining reactions

(i) *Background*

Gas–metal reactions are important methods in pyrometallurgy for the removal of impurities such as carbon and nitrogen from steel. The apparent rate of these reactions can be controlled by mass transfer of the impurities in the diffusion layer of the molten metal phase. Induction or gas stirring has been used to enhance the mass transfer.

Previous researchers have reported that Marangoni convection due to concentration gradient of surface active species enhance the apparent reaction rate. The effect of minor elements on the surface tension of liquid iron is shown in figure 2.

Lange & Wilken (1983) observed a vigorous movement on the surface of liquid iron when a low momentum jet of oxygen was blown onto the surface. They explained that this movement was Marangoni convection caused by an oxygen concentration gradient. Mizukami *et al.* (1988) studied the rate of nitrogen removal from molten steel to the gas phase by blowing a reducing gas (H_2) over the liquid surface. They found that under a reducing gas, higher apparent rate constants for nitrogen removal were obtained compared to the use of an inert gas, such as argon. They concluded that hydrogen reacted with the oxygen in the molten steel which resulted in an oxygen concentration gradient. This gave rise to Marangoni convection which enhanced the apparent rate constants. Yamaguchi *et al.* (1992) studied the effect of reductive gas blowing on the rate of the decarburization reaction under vacuum conditions. When the reducing hydrogen gas was blown over the surface of liquid iron during the decarburization process, the apparent decarburization rate constants were greater than the values obtained with argon gas blowing. Again, they attributed the greater decarburization rate constant to Marangoni convection due to the development of an oxygen concentration gradient.

In all the preceding studies, induction or gas stirring was used. Therefore, it is difficult to distinguish between Marangoni enhanced flow and flow as a result of induction or gas stirring under the reported experimental conditions. Nitrogen absorption into a low-oxygen and sulphur-containing liquid iron was chosen as the experimental system by Hirashima *et al.* (Hirashima *et al.* 1995) because mass transfer in the liquid iron phase was known to control the apparent reaction rate under this condition.

(ii) *Experimental*

A detailed description of the arrangement within the reaction tube is shown in figure 3. Liquid iron was contained in a rectangular, parallelepiped, alumina boat ($20 \times 50 \times 40 \text{ mm}^3$). Nitrogen gas was introduced through a lance with an inner diameter of 3 mm onto the iron surface. The gas flow rate was $42 \times 10^{-3} \text{ m}^3 \text{ s}^{-1}$. Argon and hydrogen gases were used for comparing the fluid motion. This reaction tube was set in a furnace with $MoSi_2$ heating elements. A X-ray system was used to obtain images of the surface movements of liquid iron by using ZrO_2 particles of 1–2 mm diameter as an inert marker. A 5 g sample of iron was taken by suction via a quartz tube which was then quenched in order to determine the nitrogen content in the iron.

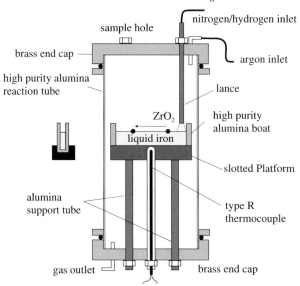

Figure 3. Schematic diagram of the experimental apparatus for nitrogen absorption.

Table 1. *Results of the measured surface motion of liquid iron*

gas	flow rate $(m^3 \, s^{-1})$	lance height (mm)	velocity $(m \, s^{-1})$	moving direction
N_2	42×10^{-3}	35	0.05 to 0.11	from lance
Ar	42×10^{-3}	35	0	—
H_2	12×10^{-3}	5	0.15 to 0.20	toward lance

(iii) *Results and discussion*

The surface velocity profile in the case of nitrogen gas blowing is shown in figure 4. The direction of flow was from the impingement area to the outer edge. A high concentration of nitrogen at the impingement area leads to a low-surface-tension zone which creates convection from a low-surface-tension area (from the lance) to a high-surface-tension area. The velocity is high in the vicinity of the nitrogen impingement area and decreases on going away from the area. The velocity varies from 0.05 to 0.11 m s^{-1}. A comparison of the flow using different gases is shown in table 1. No movement on the surface was observed for the case of Ar injection. When H_2 is used as the blowing gas, the direction of flow was opposite to that for the case of N_2 injection. A high concentration of hydrogen at the impingement area leads to a low concentration of oxygen because hydrogen reacts with oxygen in the molten iron. This results in a high-surface-tension zone. A convection was created from a region of low surface tension to a high-surface-tension area (towards the lance). From these results, it is obvious that flow is induced by the surface-tension gradient and not by the momentum of the gases.

Figure 5 shows a plot between At/V (s m^{-1}) and $\ln([\%N]^i - [\%N]_0)/([\%N]^i - [\%N]_t)$. If the apparent reaction is first order, the slope will indicate the apparent mass transfer coefficient. It was confirmed that at the experimental gas flow rate, the mass transfer coefficient was not affected. This suggests that the concentration

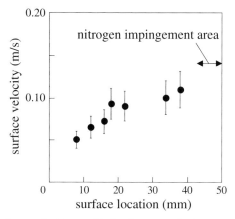

Figure 4. Surface velocity profile in the case of nitrogen gas blowing.

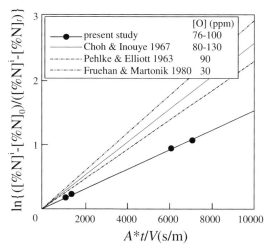

Figure 5. Comparison of the present apparent mass transfer coefficients with reported literature values for nitrogen absorption into low-oxygen and sulphur-containing liquid iron.

at the interface reaches the solubility limit of nitrogen into the iron. It would appear that the electromagnetic force was the main force to induce flow in the case of the experimental data reported in the literature. Under the condition of only Marangoni flow, the mass transfer coefficient is less than that when electromagnetic forces operate. However, because the order of magnitude of the coefficient is similar, Marangoni convection has the potential to enhance the reaction rate.

Calculations were performed assuming such experimental conditions. The results showed good agreement both for velocity and for concentration by taking turbulence into account.

(b) Electrolysis

(i) Backgound

The Hall–Heroult process is widely used to produce primary aluminium from alumina dissolved in molten cryolite by using electrolysis. A dissolution reaction shown in equation (3.1) occurs at the aluminium–bath interface where the electric field is

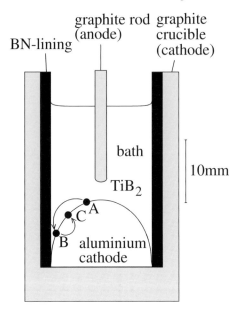

Figure 6. Experimental electrolysis cell and observed movement of particles.

small, and decreases the current efficiency:

$$Al \rightarrow Al^{3+} + 3e^-. \tag{3.1}$$

The overall rate of this reaction is generally accepted to be controlled by mass transfer of Al^{3+} into the bath phase. Thus all factors which affect the bath–metal boundary will have an influence on the current efficiency.

(ii) *Experimental*

The experimental electrolysis cell is shown in figure 6. The experiments were carried out in a graphite crucible with an inner diameter of 20 mm. The crucible was lined with boron nitride to prevent passage of current to the side of the crucible. A graphite rod with a diameter of 3 mm was used as the anode. In addition to cryolite, the bath contained 5 mass% CaF_2, 8 mass% AlF_3 and 3 mass% Al_2O_3. Small particles of TiB_2 were used as a marker at the bath–metal interface to serve as movement indicators. X-ray radiography was used to obtain the image of the particles and the interface.

(iii) *Results and discussion*

The effect of the cryolite ratio on the interfacial tension between aluminium and cryolite is shown in figure 7. Most workers have found that the interfacial tension increases with increase in AlF_3 content in the bath.

On applying a positive current (aluminium drop being the cathode) the drop apex increased, indicating an increase in the interfacial tension. The opposite effect was observed when a negative current was applied. This appeared to be an instantaneous effect which indicates that it is associated with charge effects at the interface.

As time proceeded, the effect of the current on the surface tension decreased. Figure 8 shows the change in the surface tension calculated from the observed interface geometry when the current was interrupted. An instantaneous drop in the interfacial tension occurred when the current was interrupted. High values of the interfacial

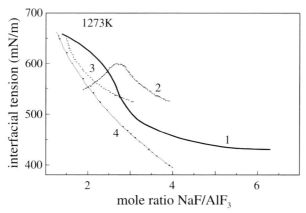

Figure 7. Interfacial tension of aluminium in cryolite melts at 1273 K: 1, Utigrad & Toguri (1985); 2, Gerasimov & Belyaev (1958); 3, Zhemchuzina & Gerasimov (1960); 4, Dewing & Desclaux (1977).

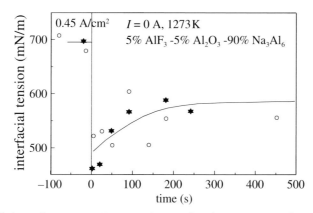

Figure 8. Effect of time after current interruption on the aluminium–cryolite interfacial tension.

tension during the electrolysis is due to the excess negative charge on the aluminium surface. After this sudden drop, the interfacial tension increases with time. This effect is caused by the accumulation of an exess amount of NaF at the interface during electrolysis (Thonstad & Rolseth 1978) and the slow diffusion of NaF away from the interface when electrolysis is stopped.

Figure 6 shows the movement of a particle placed at the bath–metal interface. Before current was applied, the particle rested in position A. Within seconds after a current of 3.1 A was applied (graphite rod anode), the particle began to move towards the wall of the crucible. It travelled the distance from A to B in approximately 1 min. The average speed was $12\,\mu\mathrm{m\,s^{-1}}$. The arrest of the particle in position B resulted because it touched the wall of the crucible. The electrolysis was interrupted after 5 min and the cell was disconnected for another 5 min before electrolysis was resumed with reverse polarity (aluminium positive). Movement of the particle was detected after 12 s. It started to move towards position C. This position was reached in 28 s after the current was applied, which leads to an average speed of $13\,\mu\mathrm{m\,s^{-1}}$.

The movement was from a region of high to low current density under conditions of 'normal' electrolysis. The boundary layer in the high-current-density region becomes more enriched with NaF than in the region with low current density due to the high

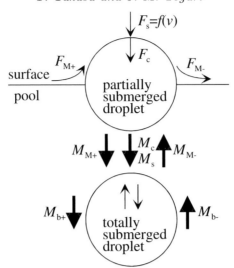

Figure 9. Forces and motions induced when a droplet falls onto a liquid pool.

transference number of sodium ions in cryolite melts. This results in an interfacial tension gradient which drives the interface from the centre to the wall and causes the TiB$_2$ particle to move down the curved interface. On the other hand, when the 'reverse' current was imposed, the high-current-density region close to the graphite electrode is more depleted in sodium ions and more enriched with aluminium ions than in the low-current-density region closer to the wall of the crucible. Then the particles moved from the wall towards the centre.

The velocity of the particles was rather low mainly due to the small current-density gradient in this system. Calculation for the pratical electrolysis conditions showed an interfacial velocity of 47 mm s^{-1} and a current loss of about 2% due to Marangoni flow.

(c) Inclusion distribution in welding

(i) *Backgound*

The importance of Marangoni flow has been widely investigated in gas–tungsten arc welding (Heiple & Roper 1982; Walsh & Savage 1985; Oreper et al. 1983; Kou & Wang 1985; Zacharia et al. 1989). When a filler metal is used, such as in gas–metal arc welding, the Marangoni interaction of the molten filler metal when it touches the weld pool has not been addressed. When there is a difference in the surface tension between filler metal and weld pool, Marangoni flow is important because the surface tension varies when the droplet hits the pool. A water–alcohol physical model was built to study the relative strengths of the driving forces responsible for fluid motion when a liquid droplet falls onto an isothermal liquid bath. There are four forces that are responsible for driving the fluid flow motion in the pool as shown in figure 9: (1) 'stirring' force, F_s, due to the momentum of the free falling droplet onto the pool; (2) 'curvature' force, F_c, due to the component of surface tension normal to the surface; (3) buoyancy force (F_{b+} or F_{b-}) due to the density difference between the droplet and pool; and (4) Marangoni force (F_{M+} or F_{M-}) due to the surface tension difference between the droplet and pool.

Table 2. *Composition of droplets and bulk liquid and their physical properties*

case	droplet	pool	$\sigma_d - \sigma_p$ (mN m^{-1})	$\rho_d - \rho_p$ (kg m^{-3})
1	water	water	0	0
2	3% alcohol	8% alcohol	9.9	8.1
3	water	8% alcohol	22.3	13.7
4	8% alcohol	3% alcohol	−9.9	−8.1
5	34% alcohol	water	−41.0	−53.9
6	water	7.6% NaCl	−2.5	−53.8

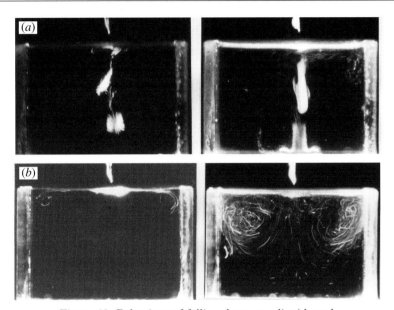

Figure 10. Behaviour of falling drop on a liquid pool.

(ii) *Experimental*

Physical model experiments using water and ethyl alcohol solution were conducted. Flow in the pool was observed by a particle tracking technique using a dark-field illumination method. An acrylic cell, 40 mm long, 20 mm wide and 30 mm high, served as the pool container. Droplets were created by means of a syringe-needle arrangement located 3 mm from the surface. The particles employed for flow tracking were hollow glass spheres (mean size 100 μm diameter).

Table 2 shows the experimental conditions employed and the cases studied.

(iii) *Results and discussion*

In case 1, at $\frac{1}{8}$ s after impact, the drop penetrated up to one-third the depth of the pool. There is no definitive flow being developed in the pool after 1 s. Hence the effects of the stirring force and curvature force on flow development are weak in this experiment.

Figure 10 shows the behaviour of a falling drop on a pool for cases 2 and 3. The droplet penetrates deep into the bath and hits the bottom of the cell within 1 s. The flow in case 3 is much more intense than in case 2. In order to determine the relative strength of the buoyancy force, the same volume of fluid was introduced slowly into the acrylic cell below the surface of the pool. It was observed that the downward flow velocities due to the buoyancy force were about 30 times smaller than those observed in case 3. Thus, it was ascertained that the deeper penetrating flow in cases 2 and 3 were due primarily to Marangoni forces.

In cases 4 and 5, radially outward surface flows were generated. Definitive toroidal flow was also observed in the bulk fluid as shown in figure 10.

In case 6, the droplet showed about same penetration depth as case 1, but the initial spreading on the surface is not evident. No definitive flow is noticeble at 1 s. Hence the flow observed in cases 4 and 5 is mainly created by the Marangoni effect.

If the surface tension of the droplet is higher than that of the pool, the droplet will penetrate into the pool and the inclusions will segregate at the bottom. On the other hand, if the surface tension of the droplet is lower than that of the pool, the droplet will spread out on the surface and inclusions will segregate near the surface of the weldment.

(d) Separation of metal or matte from slag

(i) Backgound

During the smelting of metal-sulphide concentrates, a two-phase liquid system consisiting of a matte phase and a slag phase is produced. The degree of separation between the two phases is one of the main factors which determines the metal losses of the process. The pay metals (Ni, Cu and Co) are generally lost to the slag phase in two forms: either as dissolved metals or as physically entrapped metal or metal-sulphide droplets. In the majority of cases, the physically entrapped inclusions are the dominant form of pay metal losses.

Two methods of slag cleaning have been employed for reducing pay metal losses to the slag. The first method involves the slow cooling of the slag followed by grinding and concentration of the copper-containing components by flotation. In the second method, the matte and slag are kept molten under reducing conditions in a furnace to allow the heavier sulphide inclusions to settle gravimetrically through the reduced slag into the matte phase. However, there is a limitation to the droplet size when separation is governed by Stokes law: a 0.5 mm diameter matte droplet settles at a rate of about 2 mm s^{-1}, but a 0.1 mm droplet will settle at about 0.07 mm s^{-1}. It is believed that the application of an electric field might assist in the separation process and thus enhance metal recovery.

The electrocapillary phenomenon which moves the droplet is caused by a difference in the interfacial tension due to the electric potential. Figure 11 shows the typical fundamental relationship between the capillary motion and the electrocapillary diagram. If the polarization is ideal (no decomposition) and the solution is of fixed condition, the Lipmann equation gives the relationship between the surface excess charge density and the interfacial tension gradient with respect to the electric potential:

$$\left(\frac{\partial \sigma}{\partial \phi}\right)_\mu = -q_m. \tag{3.2}$$

At a potential of zero, $\partial \sigma / \partial \phi$ has a negative value because the metal surface is usually positively charged when no field is applied. A high value of the absolute potential decreases the interfacial tension since the charges repel each other and then reduce

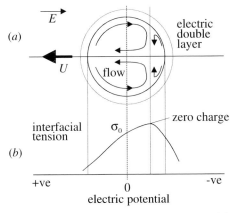

Figure 11. Relationship between (*a*) electrocapillary motion and (*b*) electrocapillary diagram.

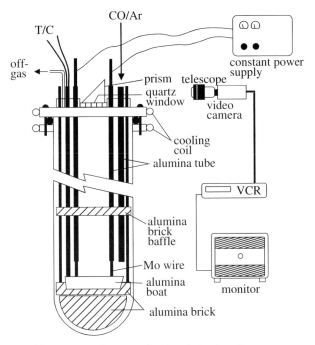

Figure 12. Experimental set-up for droplet migration measurements.

the surface tension. When an electric field is imposed on the droplet, the interfacial tension has a maximum value at the location where the potential corresponds to the maximum in the electrocapillary diagram. Then, two circulating flows are expected. This surface motion of the droplet will also drag adjacent layers of electrolyte. As a result, the entire droplet moves. The direction of the movement depends on the dominant flow which is created by the steeper slope in the electocapillary diagram.

(ii) *Experimental*

Figure 12 is a schematic representation of the experimental set-up. An alumina boat (60 mm by 15 mm) was used as a container for the working slag. A pre-mixed gas mixture of 50% Ar–CO was introduced into the alumina reaction tube. Mo wire electrodes are located at both ends of the boat. A single particle of previously weighed

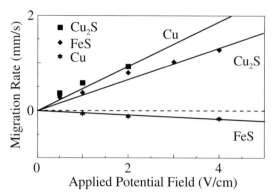

Figure 13. Effect of applied electric field on the migration rates of Cu, Cu$_2$S (80%Cu) and FeS droplets on the surface of synthetic fayalite slag (70%FeO) at 1523 K.

sulphide was introduced onto the slag surface via a quartz tube. When the particle had melted, which was almost instantaneously, it was kept on the slag surface by surface-tension forces. The droplet migration was observed from above.

(iii) *Results and discussion*

Levich (1962) has derived an equation for the migration rate as shown in equation (3.3):

$$U = \frac{\epsilon E a}{(2\mu + 3\mu')(1 + (a/2\chi\omega)) + (\epsilon^2/\chi)},$$
(3.3)

$$\omega = \frac{RT}{zFi_{\text{lim}}},$$
(3.4)

where ϵ (C m^{-2}) is the surface excess charge density, E (V m^{-1}) is the electric field, a (m) is the droplet radius, μ (Pa s) is the viscosity of the electrolyte, μ' (Pa s) is the viscosity of the droplet, χ (S m^{-1}) is the electrical conductivity of the electrolyte, ω (Ω m^{-2}) is the interfacial resistance, R (j K^{-1} mol^{-1}) is the gas constant, T (K) is temperature, z is valency, F (C mol^{-1}) is Faraday's constant and i_{lim} (A m^{-2}) is the limiting current density. The term relating to ω in equation (3.3) means that the redox reaction will occur at the interface and the droplet is no longer ideally polarized if the potential exceeds the decomposition potential of the chemical species in the droplet.

Figure 13 shows the migration rates of Cu$_2$S, FeS and Cu droplets on synthetic fayalite slag. The positive velocity denotes migration to the anode. The migration rate increased with increase in the applied electric field. The direction of migration depends on the material of the droplet. The migration rates of solid Ni, Fe, C and Al$_2$O$_3$ were also investigated and these particles remain almost motionless. The intensity of the electrocapillary motion is thought to be higher than that of electrophoresis.

Figure 14 shows the effect of droplet size on the migration rates of Cu$_2$S droplets on the surface of synthetic fayalite slag. Increase in the droplet size increases the migration rate. This is because the Marangoni force is proportional to the surface area while the viscosity force is proportional to the radius. However, the effect shows the tendency to saturate in the case of larger radius. This effect is thought to be caused by an increase in the decomposition, as expressed in equation (3.3).

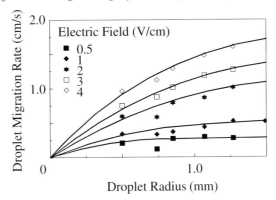

Figure 14. Effect of droplet size on the migration rates of Cu$_2$S droplet on the surface of synthetic fayalite slag (70%FeO) at 1523 K.

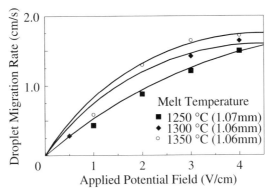

Figure 15. Effect of applied electric field on the migration rates of Cu$_2$S droplets on the surface of synthetic fayalite slag (70%FeO) at various melt temperatures.

Figure 15 shows the effect of temperature on the migration rate. The migration rate increases with increase in temperature. This is in accordance with equation (3.3) where the higher temperatures result in lower slag viscosities and higher electrical conductivities.

4. Summary

The mechanism and contribution of Marangoni flow to the following pyrometallurgical processes are discussed: the effect of changing the nitrogen composition on the surface of liquid iron, current efficiency in aluminium electrolysis, inclusion distribution in gas–metal arc welding, and the separation of metal or matte droplets from slag. It was shown that the Marangoni flow affects the mass transport of dissolved materials, inclusions or droplets. An understanding of the role of Marangoni flow on the kinetics of these operations is important. Further effort should be directed to investigate whether Marangoni convection occurs in current industrial processes in order to advantageously maximize the use of this phenomenon and also to develop new processes based on the application of Marangoni flow.

References

Choh, T. & Inouye, M. 1967 *Tetsu-to-Hagane* **53**, 1393-1406.

Choo, R. T. C., Warczok, A. & Toguri, J. M. 1991 Experimental studies of the electrodynamic behaviour of metal and metal sulphide droplets in slag. In *Pyrometallurgy of Copper, Copper 91-Cobre 91* (ed. C. Diaz, C. Landolt, A. Luraschi & C. J. Newman), vol. 4, pp. 409–424. CIM, Pergamon Press.

Choo, R. T. C. & Toguri, J. M. 1992 *Canadian Metall. Q.* **31**, 113–126.

Choo, R. T. C., Mukai, K. & Toguri, J. M. 1992 *Welding Res. Suppl.* **71**, 139s–146s.

Dewing, E. W. & Desclaux, P. 1977 *Met. Trans.* B **8**, 555–561.

Fruehan, R. J. & Martonik, L. J. 1980 *Tetsu-to-Hagane* B **11**, 615–621.

Gerasimov, A. D. & Belyaev, A. I. 1958 *Izv. Vyssh. Uchebn. Zaved, Tsvet. Met.* **1**, 58.

Heiple, C. R. & Roper, J. R. 1982, *Welding J.* **61**, 97s–102s.

Hirashima, N., Choo, R. T. C., Toguri, J. M. & Mukai, K. 1995 *Metall. Mater. Trans.* B **26**, 971–980.

Itoh, S., Choo, R. T. C. & Toguri, J. M. 1995 *Canadian Metall. Q.* **34**, 319–330.

Kou, S. & Wang, Y. H. 1985 *Welding J.* **65**, 63s–70s.

Kozakevitch, P. & Urbain, G. 1961 *Mem. Sci. Rev. Metall.* **58**, 931–947.

Lange, K. W. & Wilken, M. 1983 *Can. Metall. Q.* **22**, 321–326.

Levich, V. G. 1962 *Physicochemical hydrodynamics*, pp. 472–531. Englewood Cliffs, NJ: Prentice-Hall.

Mizukami, Y., Mukawa, S., Saeki, T., Shima, H., Onoyama, S., Komai, T. & Takaishi, S. 1988 *Tetsu-to-hagane* **74**, 294–301.

Oreper, G. M., Eagar, T. W. & Szekely, J. 1983 *Welding J.* **62**, 307s–312s.

Pehlke, R. D. & Elliott, J. F. 1963 *Trans. AIME* **227**, 844–855.

Thonstad, J. & Rolseth, S. 1978 *Electrochim. Acta* **23**, 223–241.

Utigard, T. & Toguri, J. M. 1985 *Met. Trans.* B **16**, 333–338.

Utigard, T. & Toguri, J. M. 1986 *Met. Trans.* B **17**, 547–552.

Utigard, T. & Toguri, J. M. 1990 Marangoni flow in the Hall–Heroult cell. In *Light Metals 1991* (ed. E. L. Rooy), pp. 273–281. The Minerals, Metals and Materials Society.

Utigard, T., Rolseth, S., Thonstad, J. & Toguri, J. M. 1989 Interfacial phenomena in aluminum electrolysis. In *Proc. Int. Symp. on Production and Electrolysis of Light Metals, Halifax* (ed. B. Closset), pp. 189–199. Pergamon Press.

Walsh, D. W. & Savage, W. F. 1985 *Welding J.* **64**, 59s–62s.

Yamaguchi, K., Takeuchi, S. & Sakuraya, T. 1992 *CAMP-ISIJ* **5**, 1274.

Zacharia, T., David, S. A., Vitek, J. M. & Debroy, T. 1989 *Welding J.* **68**, 499s–519s.

Zhemchuzhina, E. A. & Belyaev, A. I. 1960 Interfacial tension at the boundary between liquid Al and molten salts. *Fiz. Kim. Rasplav. Soleii Shlakov, Akad Nauk SSSR, Uralsk Filial, Inst. Elextrochim., Tr. Vses. Soveshch., Sverdlovsk*, pp. 207–214.

Direct observation of spontaneous emulsification and associated interfacial phenomena at the slag–steel interface

By Yongsug Chung and Alan W. Cramb

Department of Materials Science and Engineering, Carnegie Mellon University,
Pittsburgh, PA 15213, USA

Interfacial tension decreases drastically when an intense chemical reaction occurs at a steel–slag interface. This phenomenon results in spontaneous droplet spreading during the initial period of reaction and droplet recovery as the rate of reaction decreases. In the present work, spreading tendency was found to be associated with spontaneous emulsification of both steel in slag and slag in steel. Spontaneous emulsification was observed at 1550 °C when a liquid Fe–3.28%Al alloy droplet was placed in contact with liquid CaO–SiO₂–Al₂O₃ (40:40:20 by weight) slag.

Surface turbulence induced metal emulsification and droplet spreading was observed by X-ray photography. Spontaneous droplet spreading was noted at alloy aluminium contents as low as 0.25%. Spontaneous emulsification was steel into slag was documented at aluminium contents greater than 3%. From the observation of quenched Fe–3.28%Al alloy droplets by optical and scanning electron microscopy, slag entrapment and metal emulsification were documented and the metal–slag interface was shown to be extremely perturbed during the reaction of aluminium with silica.

Keywords: X-ray photography; spontaneous emulsification; interfacial tension;
steel–slag interface; casting

1. Introduction

When a chemical reaction takes place between a droplet of a liquid steel alloy and a liquid slag a number of investigators (Kozakevitch *et al.* 1955; Ooi *et al.* 1974; Riboud & Lucas 1981; Sharan & Cramb 1995; Liukkonen *et al.* 1997; Gaye *et al.* 1984) have found that the droplet spontaneously spreads on its substrate and then, after an amount of time, spontaneously recovers its equilibrium shape. Most investigators have calculated the apparent interfacial energy of the droplet during spreading by assuming that the Young–Laplace equation still holds during spreading and that the droplet remains symmetrical. Under these assumptions it appears that the interfacial energy is decreasing rapidly during reaction.

This phenomenon may be important when aluminium is added to a molten steel in contact with silicate slag, a common occurrence during either steel deoxidation or slag 'killing' (when aluminium is added in bulk to the slag covering on top of a steel ladle to reduce the iron and manganese oxide content of the slag). Few studies of dynamic interfacial phenomena have been carried out on steel alloys with less than 4% aluminium in contact with silicate slags. In this study, observation of dynamic interfacial phenomena was carried out with the aid of X-ray photography at a range

Table 1. *Reactions for which a drastic decrease of interfacial tension has been observed (Riboud & Lucas 1981)*

metallic alloys	slag	reaction
Fe–Al	$CaO–SiO_2$ (or $–Al_2O_3$)	$2\underline{Al} + \frac{3}{2}SiO_2 = Al_2O_3 + \frac{3}{2}\underline{Si}$
Fe–Al	$CaO–Al_2O_3–Fe_2O_3$	$2\underline{Al} + Fe_2O_3 = Al_2O_3 + 2\underline{Si}$
Fe–C–S	$CaO–Al_2O_3–SiO_2$	$\underline{S} + (O) = (S) + \underline{O}$
Fe–Ti	$CaO–Al_2O_3–SiO_2$	$\underline{Ti} + SiO_2 = TiO_2 + Si$
Fe–P	$CaO–Al_2O_3–Fe_2O_3$	$2\underline{P} + \frac{5}{3}Fe_2O_3 = P_2O_5 + \frac{10}{3}Fe$
Fe–B	$CaO–Al_2O_3–SiO_2–Fe_2O_3$	$2\underline{B} + Fe_2O_3 = B_2O_3 + 2Fe$
Fe–Cr	$CaO–SiO_2–FeO$	$2\underline{Cr} + 3FeO = Cr_2O_3 + 3Fe$
Fe	Cu_2O	$Fe + Cu_2O = 2\underline{Cu} + FeO$
Fe–Si	$Cu_2O–Al_2O_3$	$\underline{Si} + 2Cu_2O = SiO_2 + 4\underline{Cu}$

of aluminium contents (0.25–3.28%) in steel alloy droplets in contact with $CaO–SiO_2–Al_2O_3$ (40:40:20 by weight) at 1550 °C.

Dynamic phenomena (droplet spreading and recovery) was documented at aluminium contents as low as 0.25% and clear evidence of spontaneous emulsification was seen in steel alloys containing greater than 3% aluminium.

2. Background

Interfacial tension (or energy), an equilibrium thermodynamic quantity, has been determined by many investigators (Gaye *et al.* 1984; Ogino *et al.* 1984; Kawai *et al.* 1982; Gammal & Schoneberg 1992; Jimbo & Cramb 1992; Liu *et al.* 1993; Jimbo *et al.* 1995, 1996), and ranges from 300 to 1700 mN m^{-1} when steels are in contact with slags. The lowest values are found when oxygen saturated iron is in contact with an FeO containing slag (Ogino *et al.* 1984; Kawau *et al.* 1982). Apparent interfacial tensions, calculated during droplet spreading, decrease to lower than 100 mN m^{-1} where a reaction between an element in the alloy and a slag component occurs. Since Kozakevitch *et al.* (1955) first reported this dynamic interfacial phenomenon between an iron, carbon and sulphur alloy and a blast furnace type slag in 1955, several investigators have reported dynamic droplet spreading in steel–slag systems (Ooi *et al.* 1974; Riboud & Lucas 1981; Sharan & Cramb 1995; Liukkonen *et al.* 1997). Riboud & Lucas (1981) measured the apparent interfacial tension between a number of metallic alloys in contact with slag systems. They observed the lowering of interfacial tension for the reactions between metallic alloys and slags shown in table 1.

Recently the effect of sulfur transfer has been shown to affect results of equilibrium interfacial energy determinations even when no droplet spreading can be determined (Jimbo *et al.* 1995, 1996).

Due to its industrial significance, a number of investigators have focused on the dynamic phenomena that can be observed when an liquid Fe–Al alloy droplet is in contact with a silicate slag. Figure 1 shows the summarized results of previous investigators (Ooi *et al.* 1974; Riboud & Lucas 1981; Sharan & Cramb 1995; Liukkonen *et al.* 1997). As can be seen in figure 1, three distinct periods can be observed: (1)

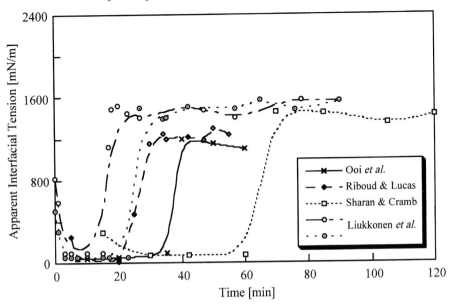

Figure 1. Dynamic interfacial phenomena observed by previous investigators.

an initial transient period where the droplet dynamically spreads and the apparent interfacial energy decreases; (2) a period of very low interfacial energy where droplet spreading stops, but the droplet remains flattened; and (3) a recovery period where the droplet returns to its equilibrium shape. In most cases, the transient period finishes within 5 min. The duration time of the period of low interfacial energy ranges from 15 min up to 1 h and during this time apparent interfacial tension values were measured to be less than $100 \ \mathrm{mN \ m^{-1}}$. The recovery period has a different start time depending on the study; however, in all cases the recovering rate was very rapid (less than 5 min). Ooi *et al.* (1974) were the first to observe spontaneous droplet spreading and an apparent lowering of the interfacial tension for Fe–Al alloy droplets containing more than 4% initial aluminium content in contact with CaO–SiO_2–Al_2O_3 slags. They determined that an aluminium content of greater than 2% was necessary to induce dynamic interfacial phenomena. They also suggested that droplet dynamic phenomena were due to the reduction of silica in the slag by aluminium dissolved in the iron and that the rate controlling step was the chemical reaction at the interface. They did not observe dynamic phenomena between an Fe–Al droplet containing 9.8% aluminium and a CaO–Al_2O_3 slag which indicated that the following reaction was responsible for the dynamic effects:

$$4\underline{Al} + 3(SiO_2) = 3\underline{Si} + 2(Al_2O_3) \tag{2.1}$$

Riboud & Lucas (1981) also reported droplet spreading between Fe–4.45%Al alloys and a CaO–SiO_2–Al_2O_3 slag and suggested that the apparent interfacial tension tends to zero when the oxygen flux is larger than about $0.1 \ \mathrm{g \ atom \ m^{-2} \ s^{-1}}$. Under this condition, they also suggested that dispersion of one phase into the other may occur spontaneously. By microscopic observation after quenching of a capillary slag–metal interface, the following features were observed by Riboud & Lucas (1981).

(1) The metal–slag interface had a very irregular outline with roughness of all sizes from millimeters down the limit of the resolution of the optical microscope (micrometres).

(2) Numerous metallic droplets were found in the slag phase near the metal–slag interface. Droplet sizes ranged from 1–100 μm in diameter. For the metallic droplet formation, they suggested a 'diffusion and stranding' mechanism.

Sharan & Cramb (1995) confirmed these phenomena in iron based alloys containing 20% nickel and 2.39% aluminium in contact with a CaO–SiO_2–Al_2O_3 (40:40:20 by weight) slag. The interfacial tension within 15 min was measured to be 296 mN m^{-1}, between 15 and 60 min it decreased to be less than 100 mN m^{-1}, and then significantly recovered to a value of about 1500 mN m^{-1} after 1 h, when equilibrium between the metal and slag was attained. As can be seen in figure 1, the long transient period followed by an even longer period of lowered interfacial energy may result from the nickel effect and/or the lower content of aluminium. Sharan suggested that the slag phase surrounding the metal droplet had a higher oxygen potential than the metal, which lead to an imbalance in the oxygen level in the two phases and was the driving force for the reaction.

Liukkonen *et al.* (1997) verified the above findings by using Fe-4.5% Al alloy in contact with a CaO–SiO_2–Al_2O_3 slag. They also verified the reaction between aluminium and silica by chemical analysis before and after experiments. However, they reported that droplet flattening was not observed when using MgO crucibles.

Due to these interesting results and the lack of data at lower aluminium contents, it was decided to further study this phenomena at aluminium contents ranging from 0.25–3.3% in order to determine the limits of dynamic droplet spreading.

3. Experimental method

Determination of the apparent and equilibrium interfacial tension between a liquid Fe–Al alloy and CaO–SiO_2–Al_2O_3 slag was carried out by using a combination of the sessile drop and X-ray photography techniques. After taking an X-ray photograph of the specimen droplet at 1550 °C, the image contour was digitized. The interfacial tension was determined numerically by finding a solution to Laplace's equation which was the best fit for the measured data points. The procedures for experimental and calculation methods are given elsewhere in detail (Jimbo & Cramb 1992).

(a) Sample preparation for dynamic interfacial tension measurement

Five Fe–Al master alloys were prepared to investigate the phenomena of dynamic interfacial tension. In order to prepare master alloys that have a different initial aluminium content, solid iron (99.95% purity obtained from Noah Technology) and solid aluminium (99.999% purity obtained from Alfa AESAR) were melted in contact with CaO–Al_2O_3 (1:1 by weight) slag.

Table 2 shows the chemical composition of five Fe–Al alloys analysed by LECO for oxygen and by Spectrolab S for the other elements from RIST (Research of Industrial Science Technology in Korea) before and after experiments. Master alloys were cut to be 1.5–3 g, polished and washed with acetone. The prepared specimen was contained under slag in an alumina crucible and charged into an electric furnace with Ar–CO (3:1 by volume) atmosphere at 1550 °C. The CaO–SiO_2–Al_2O_3 (40:40:20 by weight) slag (10 g) was prepared in the induction furnace with an air atmosphere. After the furnace reached 1550 °C, the crucible containing the metal–slag specimen was placed in the middle of the furnace. It took about 4 min to melt the sample after it charged, as confirmed by X-ray observation.

Table 2. *The chemical composition of alloys before and after experiments*

(Top row, before experiments; bottom row, after experiments (wt%).)

sample	Al	Si	O	S	C
1	0.25	0.03	0.024	0.002	0.001
	0.002	0.316	0.020	0.005	0.011
2	0.73	0.017	0.019	0.002	0.013
	0.002	0.536	0.015	0.006	0.052
3	0.77	0.012	0.007	<0.001	—
	< 0.001	< 0.39	0.11	—	—
4	1.84	0.036	0.012	0.003	0.041
	0.006	1.31	0.075	0.005	0.031
5	3.28	0.03	0.005	<0.001	—
	<0.01	1.53	0.13	—	—

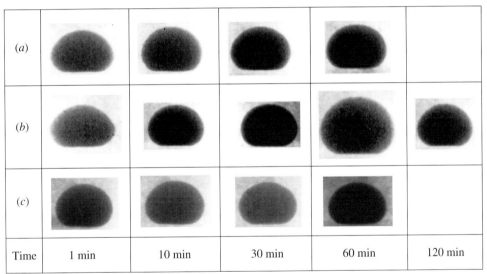

Time	1 min	10 min	30 min	60 min	120 min

Figure 2. The scanned X-ray photographs of Fe–Al alloys in contact with $CaO–SiO_2–20\%Al_2O_3$ (basicity, $B = 1{:}1$) slag as a function of time: (*a*) Fe–0.25% Al; (*b*) Fe–0.73% Al; (*c*) Fe–0.77% Al.

4. Results and discussion

(*a*) *X-ray observation and interfacial tensions*

Figure 2 shows the Fe–Al (less than 1%Al) droplet shapes after immersion into a $CaO–SiO_2–Al_2O_3$ (40:40:20 by weight) slag as a function of time. As can be seen in figure 2*a*, the droplet shape at 1 min is flatter than after 10 min and indicates that the dynamic droplet spreading occurs even when aluminium contents are as low as 0.25%. Spreading was complete and the droplet recovered in 10 min. The droplet shape remains almost unchanged between 10 min and 1 h. A similar trend can be seen in the case of the Fe–0.73% Al alloy in figure 2*b*; however, the droplet shape at 1 min is much flatter than the one for the Fe–0.25% Al alloy. In this case the photograph

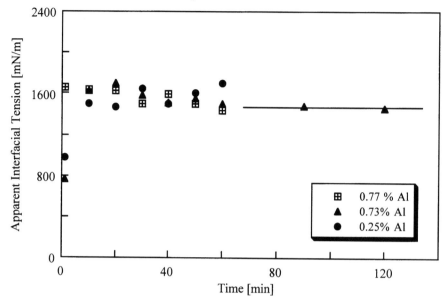

Figure 3. The variation of apparent interfacial tension between Fe–($< 1\%$)Al alloy and CaO–SiO$_2$–20%Al$_2$O$_3$ (B = 1:1) slag.

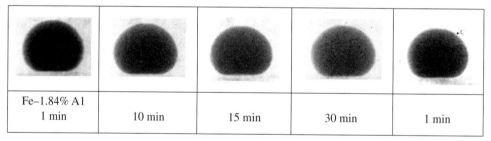

| Fe–1.84% Al 1 min | 10 min | 15 min | 30 min | 1 min |

Figure 4. The scanned photographs of Fe–1.84% Al alloys in contact with CaO–SiO$_2$–20% Al$_2$O$_3$ (B = 1:1) slag as a function of time.

was taken for 2 h. The droplet shape at 2 h is similar to the one at 10 min. In the case of 0.77% Al alloy (figure 2c), droplet spreading was not observed.

Figure 3 shows the calculated variations of apparent interfacial tension based on the droplets of figure 2. As can be expected from the observation, interfacial tensions at 1 min in two of the samples are lower than the other measurements. The interfacial tension of Fe–0.73% Al (760 mN m^{-1}) is lower than that of Fe–0.25% Al (980 mN m^{-1}). For the Fe–0.25% Al alloy, the interfacial tension values are initially scattered; however, after 30 min the scatter in the values decreased. In the case of Fe–0.73% Al alloy the interfacial tension increases initially and converges on an equilibrium value after 30 min. For this alloy the measurement was carried out for 2 h and the value of interfacial tension did not significantly vary from 30 min to 2 h.

Figure 4 shows the droplet shapes of an Fe–1.84% Al alloy in contact with CaO–SiO$_2$–Al$_2$O$_3$ slag system with time. Unlike previous cases, there was no droplet spreading observed even at 1 min.

Figure 5 shows the calculated interfacial tensions for these droplets. In the case of Fe–1.84% Al alloy, the apparent interfacial tensions do not come to an equilibrium value but increase with time. This result was quite surprising as the final values for

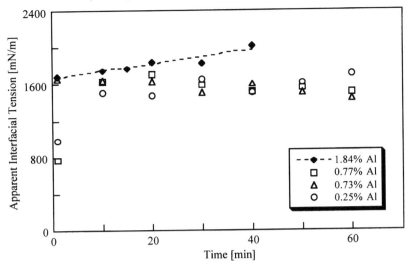

Figure 5. The variation of apparent interfacial tension between Fe–1.84% Al alloy and
CaO–SiO$_2$–20% Al$_2$O$_3$ (B = 1:1) slag (previous values shown for comparison).

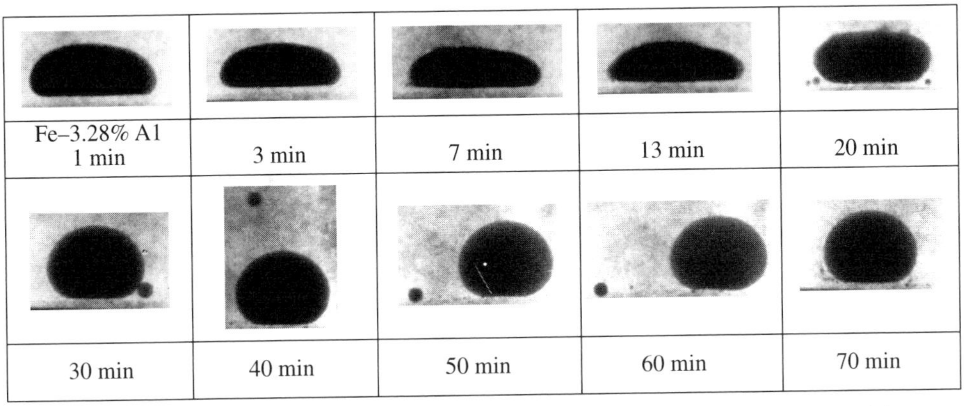

Fe–3.28% Al 1 min	3 min	7 min	13 min	20 min
30 min	40 min	50 min	60 min	70 min

Figure 6. The scanned X-ray photographs of Fe–3.28% Al alloy in contact with
CaO–SiO$_2$–20% Al$_2$O$_3$ slag as a function of time.

the apparent interfacial tension were higher than any equilibrium measurements and indicative of droplet swelling (increase in height) rather than spreading (decrease in height).

Figure 6 shows the appearance of the Fe–3.28% Al droplets as a function of time. The droplet shape at 3 min is much flatter than that at 1 min. From 7 to 13 min the droplet image is blurred especially at the left-hand side and at the interface on the top side of the droplet. The time for the exposure is 4 s. This indicates that the interface is disturbed and a vigorous reaction between the alloy and the slag occurs in this time period. At 20 min the edge becomes sharp; however, perturbation of the top interface still occurs and leads to blurring in the photograph. At this time, in addition to the main droplet, there are three small droplets that have spontaneously separated from the main droplet. This figure clearly shows spontaneous metal emulsification due to reaction between the metal and a slag.

Between 30 and 70 min the droplet shape recovered. It was also observed that the small droplet that exists between 30 and 60 min is travelling around the large

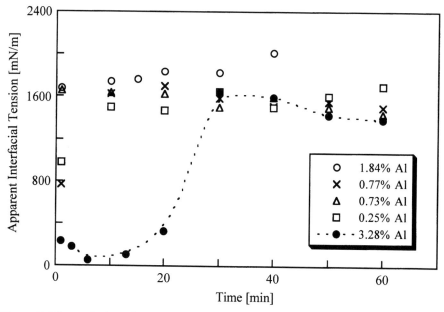

Figure 7. The variation of apparent interfacial tension between Fe–3.28% Al alloy and CaO–SiO$_2$–20% Al$_2$O$_3$ (B = 1:1) slag.

droplet. This movement indicates that fluid flow exists in the slag phase even under apparent constant temperature conditions. After 1 h only one droplet could be seen and on quenching the sample only one droplet could be found in the crucible.

Figure 7 shows the measured apparent interfacial tension values for the above system. The apparent interfacial tensions are calculated at 1 min to be 230 mN m^{-1} and at 3 min to be 180 mN m^{-1}. Between 7 and 20 min, it is difficult to capture the contour of the interface from X-ray images owing to the blurred interface. For those cases, the interfaces are selected as a smooth curved line from a clear interface and apparent values are 57, 106 and 328 mN m^{-1} at 6, 13 and 20 min, respectively. The interfacial tension after 30 min reaches an equilibrium value ranging from 1500 to 1600 mN m^{-1}. Apparent droplet densities have been calculated from X-ray photographs and their values (approximately 7000 kg m^{-3}) did not significantly change with time after 30 min, even though spontaneous emulsification can be seen in the photograph. However, at 20 min, the apparent density is calculated to be 5180 kg m^{-3} and indicates that the volume of metal has significantly expanded. Both results may be due to emulsification of both slag and metal into each other which could result in a higher apparent volume of the droplet.

The cause of dynamic interfacial phenomena during mass transfer, although well documented, is not well understood. In this discussion, basic thermodynamic and kinetic aspects will be considered. In table 2 results of chemical analysis before and after experiments is given. Silicon was picked up by the iron while aluminium was transferred into the slag and mass balance indicates that the following overall reaction occurred:

$$4\underline{Al}_{metal} + 3(SiO_2)_{slag} = 3\underline{Si}_{metal} + 2(Al_2O_3)_{slag} \qquad (4.1)$$

Figure 8 shows that the exchange relationship between the decrease in aluminium and the increase in silicon of the alloy before and after the experiments. The straight line presents the stoichiometric relationship between aluminium and silicon from

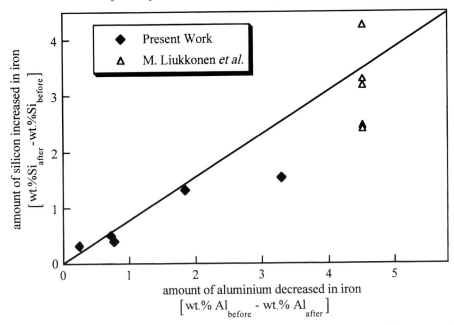

Figure 8. The relationship between the amount of aluminium decrease and the amount of silicon increase in iron from chemical analysis before and after experiments.

equation (4.1). The data qualitatively fit the line, which indicates that the overall reaction (4.1) could be responsible for the dynamic phenomena.

(b) Micro-observation after quenching

To observe the metal–slag interface during reaction, the samples were quenched at 5, 10, 20 and 25 min after taking X-ray photographs. Figure 9 shows the X-ray photographs and their calculated apparent interfacial tension values. The apparent interfacial tension at 5 min represents a higher value than the previous experiment. This may be due to a delayed initiation of the reaction. The other values are consistent with previous results. A quenched sample can allow a direct observation of the steel–slag interface. Two quenched samples at 5 and 10 min were observed using an optical and scanning electron microscope (SEM), where quenching occurred immediately after taking an X-ray photograph.

Figure 10a shows the overall shape of the alloy surrounded by slag at 5 min after quenching. As can be expected from the X-ray photographs, the droplet is almost spherical in shape. However, a deformed part was observed by optical microscopy. This part is magnified in figure 10b where entrapped slag could be seen close to the interface. Figure 11 shows an SEM image of the entrapped slag where its chemistry was verified by EDS analysis. These Al, Ca and Si mappings clearly verify that there is entrapped slag within the droplet. From the observation by optical microscopy (figure 10b), the entrapped slags are aligned normal to the interface and their lengths vary from 10 μm up to 0.4 mm and widths varying from 10 to 50 μm. The maximum penetration depth is 1.33 mm. This is believed by the authors to be the first indications of spontaneous emulsification of slag in steel during reaction.

Figure 12 shows a photograph of the quenched sample 10 min after the reaction started. Important metallurgical phenomena such as a significant surface creation and spontaneous metal and slag emulsification can be clearly observed. The overall

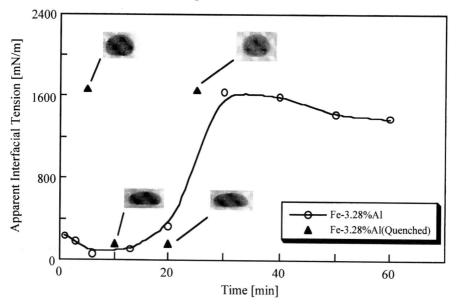

Figure 9. The variation of apparent interfacial tension between Fe–3.28% Al alloy and
CaO–SiO$_2$–20% Al$_2$O$_3$ (B = 1:1) slag.

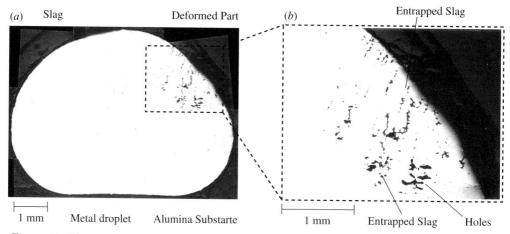

Figure 10. The appearance of the quenched Fe–3.28% Al alloy at 5 min (*a*) and the deformed
and contaminated part with higher magnification (*b*).

shape of the metal droplet looks like the picture obtained by X-ray, but the interface
has significant perturbations. Many small metal droplets (between 1 and 10 μm) as
well as large droplets (from 10 to 100 μm) were observed near the interface. Even
though both top and bottom interfaces are rough on the macroscale their features
are quite distinct on the microscale.

Figure 13 shows the images of the top interfaces ((*a*) and (*b*)) and of the bottom
interfaces ((*c*) and (*d*)) of the droplet of figure 12 observed by SEM. As can be seen
in these phonographs, there are two distinct observations:

(1) No metallic fragments can be seen near the top interface while many fragments
are found near the bottom interface (within 100 μm in size).

(2) The top interface is smooth while the bottom interface is highly perturbed.

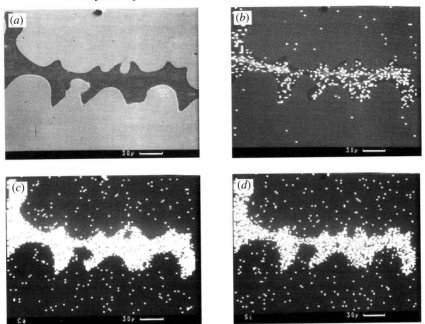

Figure 11. The SEM image of an entrapped slag and EDX analysis: (*a*) SEM image of entrapped part; (*b*) Al mapping; (*c*) Ca mapping; (*d*) Si mapping.

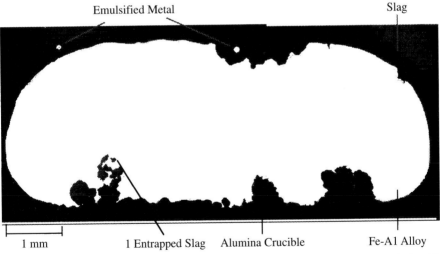

Figure 12. The photograph of Fe–3.28% Al alloy droplet in contact with CaO–SiO$_2$–Al$_2$O$_3$ quenched 10 min after melting.

It is clear that during reaction both spontaneous emulsification and droplet spreading occurs. The energy source responsible for bulk movement of slag into steel or steel into slag is not yet clear and may be related to Marangoni flows caused by local concentration gradients or local thermal gradients. The exothermic nature of aluminium reduction of silica will lead to significant thermal gradients; therefore, there will be local thermal gradients and the energy source for emulsification may be a combination of Marangoni flow by concentration and thermal gradients and natural convection.

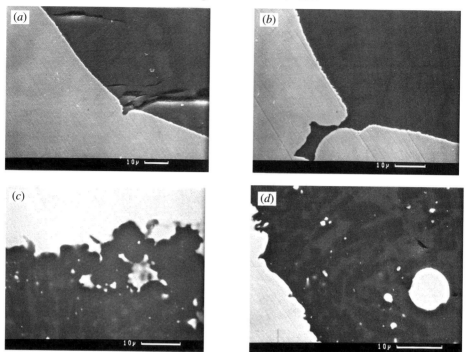

Figure 13. The features of top and bottom interfaces observed by SEM. (*a*) SEM image of top interface; (*b*) SEM image of top interface; (*c*) SEM image of low interface; (*d*) SEM image of low interface.

It is clear, however, that an understanding of dynamic phenomena can be confounded by spontaneous emulsification as droplet emulsification and swelling can be observed. Therefore, it is possible to calculate apparent interfacial energies that are lower than and also higher than equilibrium predictions depending upon the direction of emulsification.

5. Conclusions

Surface turbulence induced metal emulsification was observed by X-ray photography when an Fe–3.28% Al alloy droplet was in contact with a CaO–SiO_2–Al_2O_3 (40:40:20 by weight) slag. In addition, dynamic interfacial effects were noticed with aluminium contents as low as 0.25%. From observations using optical microscopy and SEM after quenching, both slag and metal emulsification phenomena were observed.

Dynamic interfacial phenomena, where spontaneous emulsification and droplet spreading and recovery are observed, may be more common than previously thought. Industrial processes where high levels of reactive alloys come into contact with reducible slags, for example during steel deoxidation, will be candidates for the observation of these phenomena.

Funding for this work was made possible by the member companies of the Center for Iron and Steelmaking research at Carnegie Mellon University.

References

Gammal, T. E. & Schoneberg, U. 1992 *Stahl Eisen* **112**, 45–49.

Gaye, H., Lucas, L. D., Olette, M. & Riboud, P. V. 1984 *Can. Metall. Q.* **23**, 179–191.

Jimbo, I. & Cramb, A. W. 1992 *ISIJ Int.* **32**, 26–35.

Jimbo, I., Sharan, A. & Cramb, A. W. 1995 *Trans. ISS* **16**, 45–52.

Jimbo, I., Chung, Y. & Cramb, A. W. 1996 *ISIJ Int.* **36**, 1–6.

Kawai, Y., Shinozaki, N. & Mori, K. 1982 *Can. Metall. Q.* **21**, 385–391.

Kozakevitch, D., Urbain, G. & Sage, M. 1955 *Rev. Metallurg.* **52**, 161.

Liu, W., Themelis, N. J., Gammal, T. & Zhao, F. L. 1993 *ISS Trans.* **14**, 39–44.

Liukkonen, M., Holappa, L. & Mu, D. 1997 *5rd Int. Conf. on Molten Slags, Fluxes and Salts, 1997, Sydney*, pp. 149–156. Iron and Steel Society.

Ogino, K., Hara, S., Miwa, T. & Kimoto, S. 1984 *Trans. ISIJ* **24**, 522–530.

Ooi, H., Nozaki, T. & Yoshii, Y. 1974 *Trans. ISIJ* **14**, 9.

Riboud, P. V. & Lucas, L. D. 1981 *Can. Metall. Q.* **20**, 199–208.

Sharan, A. & Cramb, A. W. 1995 *Metall. Trans.* B **26**, 87–94.

Discussion

A. PASSERONE (*ICTAM-CNR, Genoa, Italy*). With reference to the instabilities Professor Cramb has shown to appear at the liquid-iron–slag interface, which can sometimes destroy the drop axial-symmetric geometry, I would like to point out that recent studies in my laboratory on the interfacial tension evolution with tin in the presence of mass transfer across the interface, have demonstrated that deep interfacial tension minima can appear, depending on the conditions of the system.

I would like to suggest that this kind of study is taken into consideration when explaining the phenomena reported here in metal–slag systems.

A. W. CRAMB. We have also seen deep minima in interfacial tension at high rates of reaction between the droplet and the slag phase and that during this condition it is possible to observe spontaneous emulsification of both phases. These minima are shown in figures 7 and 9.

Interfacial tension effects on slag–metal reactions

By A. Jakobsson[1], M. Nasu[2], J. Mangwiru[2], K. C. Mills[2]
and S. Seetharaman[1]

[1] *Division of Theoretical Metallurgy, Royal Institute of Technology,*
SE-100 44 Stockholm, Sweden
[2] *Department of Materials, Imperial College of Science, Technology and Medicine,*
London SW7 2BP, UK

Dynamic X-ray imaging of a metal droplet in a slag phase is an elegant way to monitor the rate of the interfacial reactions. In an attempt to follow the kinetics of dephosphorization, the changes in the shape of the sessile drop of molten iron with 0.1 wt% P immersed in a slag kept in MgO crucibles were recorded as a function of time. The slag had an initial composition 40 wt% CaO, 30 wt% SiO_2 and 30 wt% Fe_2O_3. While in a set of runs, solid iron was added to the molten slag, the sequence was reversed in others, where a pellet of the slag was added to the molten iron. In the former case, it was found that after 10 s, the drop flattened and the contact angle, θ was well below $90°$, indicating a significant increase in the contact area. Beyond 250 s, the contact angle showed a gradual increase. In the second series, where the slag was added to the metal, the contact angle remained high throughout the dephosphorization reaction, indicating a rapid mass transfer of phosphorus.

Similar application in the field of copper metallurgy, where the transfer of arsenic and antimony from copper metal to molten Na_2CO_3 was studied, confirm that this technique is very useful in understanding the mechanisms of the slag–metal reactions.

Keywords: Marangoni flow; surface tension; contact angle;
sessile drop; slag–metal reactions; metal refining

1. Introduction

In modern processes, such as bath smelting and basic oxygen steelmaking, fast reaction kinetics are obtained by creating an emulsion of metal in the slag phase, leading to a large surface-area/mass ratio. In experiments carried out at the beginning of the century on the formation of emulsions, NaOH was injected onto opposite sides of a drop of gear and rape seed oil. The NaOH saponified the oil and the mass transfer resulted in, sequentially, a low interfacial tension, large surface tension gradients around the drop, Marangoni–Behara flows, turbulance and disintegration of the oil globule into a fine emulsion. Work carried out by Kozakevitch (1957) indicated that the desulphurization of iron was accompanied by a rapid decrease in the slag–metal interfacial tension. Subsequent results obtained by Gaye *et al.* (1984) indicated that for slag–metal reactions, rapid mass transfer was accompanied by a sharp decrease in interfacial tension γ_{MS} to a very low value and when the mass transfer gradually decreased, the interfacial tension increased to its original value. Obviously, a low interfacial tension would encourage emulsification and faster kinetics. It was our

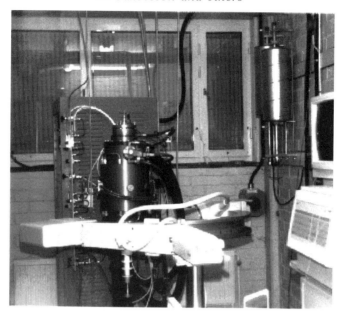

Figure 1. The sessile-drop unit.

aim in this investigation to see what effects the interfacial characteristics had on the kinetics of certain slag–metal reactions:

(i) the dephosphorization of iron to see if P, which is less surface active than S, would have the same effect on γ_{MS}; and

(ii) the removal of As and Sb from liquid copper to see if the dramatic decrease in γ_{MS} also occurred in non-ferrous systems.

2. Experimental aspects

The X-ray image analyser used for the dynamic interfacial tension studies is shown in figure 1. The X-ray unit is a Philips BV-26 imaging system with an X-ray source of 40-105 kV. The X-ray tube has a focal spot size that varies between 0.6 and 1.5 mm. The imaging system consists of a CCD camera, with a digital noise-reduction. The unit is capable of detecting dynamic movements and has a storage capacity of 34 images. The recording system consists of an IBM-PC equipped with a video card to monitor and record the X-ray images at a maximum rate of 12 pictures per second. The graphics have a resolution of 575×900 matrix. The system can store 192 images with a video memory of 32 images.

The furnace used in the equipment was acquired from Thermal Technology Inc. (model 1000-3500-FP20). It is equipped with graphite heating elements (effect = 20 kVA) and is capable of attaining a maximum temperature of 2300 °C. The furnace is controlled by a Raytek single-colour optical pyrometer. The outer jacket of the furnace is provided with quartz windows of 40 mm diameter on either side for the X-rays from the source and the detector. The reaction tube inside the furnace is made of recrystallized Al_2O_3. Appropriate radiation shields are provided to ensure a good even-temperature zone, the length of which is about 100 mm in the middle of the furnace. The assembly is capable of operating under vacuum, inert gas or gas mixtures. Appropriate gas-cleaning trains have been assembled to ensure that

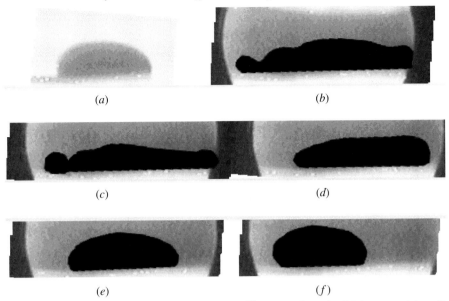

Figure 2. X-ray images of an iron drop with 0.1 wt% P in a slag of initial composition 40 wt% CaO–30 wt% SiO$_2$–30 wt% Fe$_2$O$_3$ in an MgO crucible at various time intervals (solid slag was added to liquid metal).

the impurity levels are very small. A ZrO$_2$–CaO oxygen probe is mounted at the outlet side of the experimental unit in order to monitor the oxygen partial pressures continuously.

3. Study of dephosphorization reactions (Nasu *et al.* 1997)

(*a*) *Experimental procedure*

The dephosphorization studies were carried out by following the change in shape of a sessile drop of molten iron containing 0.1 wt% P. The slag used had the initial composition of 40 wt.% CaO + 30 wt% SiO$_2$ + 30 wt% Fe$_2$O$_3$. The choice of the crucibles was limited; zirconia, as well as thoria, crucibles could not be used as they were opaque to X-rays. With alumina crucibles, Al$_2$O$_3$ pick-up was excessive. Some preliminary experiments were carried out with boron nitride crucibles, but the results had to be discarded due to the low dephosphorization levels, presumably due to the reaction with the crucible forming a silico-borate slag. The only choice of crucible material was MgO and three experiments were performed with MgO crucibles. The experiments were carried out in two different ways:

(a) addition of a slag-compact to the molten metal; and
(b) addition of solid metal to the liquid slag.

(*b*) *Results and discussion*

(i) *Addition of solid slag to the liquid metal*

Figure 2*a* shows the initial molten Fe–P alloy just before the slag was added. It is seen that the contact angle between the molten metal drop and the MgO base of the crucible is greater than 90°. With the gas atmosphere prevailing over the metal, the metal is non-wetting with respect to the crucibles. At 10 s the drop has flattened and θ is quite low ($\theta \ll 90°$), as can be seen in figure 2*b*. As the slag replaces the

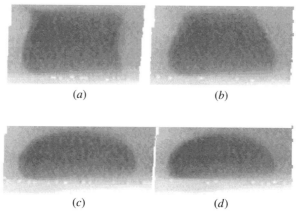

Figure 3. X-ray images of an iron drop with 0.1 wt% P in a slag of initial composition 40 wt% CaO–30 wt% SiO_2–30 wt% Fe_2O_3 in an MgO crucible at various time intervals (solid metal was added to liquid slag).

gas atmosphere surrounding the metal droplet, there is a significant drop in the interfacial tension (γ_{MS}) between the metal and the slag. As mentioned earlier, this would lead to an increase in the contact area between the two phases with high mass-transfer conditions. These conditions favour emulsification of the metal drop in the slag. This is evident at higher reaction times when the metal drop develops a neck. This situation is illustrated in figures 2c, d, corresponding to 70 and 231 s, respectively. Even slight disturbances at this stage can lead to the breaking of the drop at the neck thus increasing the interfacial area further. However, beyond 278 s, the shape of the drop changes resulting in an increase in the contact angle and the interfacial tension. It can be expected that the intense mass-transfer step may have slowed down very much and the system is near equilibrium conditions.

The initial rapid decrease in the interfacial tension during a rapid mass-transfer step followed by a gradual increase in the interfacial tension is similar to the interfacial phenomenon changes observed in the case of desulphurization. Kozakevitch (1957) has shown that due to the transfer of sulphur from a drop of iron containing carbon and sulphur to a CaO–Al_2O_3–SiO_2 slag at 1550 °C, the interfacial tensions may even decrease to 5 dyne cm^{-1}. The reason for this drop is attributed to the simultaneous transfer of Fe and S to the slag phase as Fe^{2+} and S^{2-} ions, providing probably a good bonding between the metal and slag phases in this process. In order to visualize a similar mechanism in the case of the transfer of P from iron metal to the slag, further experimentation may be necessary.

(ii) *Addition of solid metal to liquid slag*

Figure 3a shows the solid metal added to the molten slag. The contours of the unmelted metal pellet are clearly seen in this figure. After 30 s, the metal phase is molten (figure 3b). The contact angle between the metal and the slag is high and so consequently is the interfacial tension. This situation remains the same throughout (figures 3c–f). This could perhaps be indicative of the end of the rapid mass-transfer step within the first 30 s, even as the solid metal melts forming the drop. This would mean that the mass transfer of P in molten iron is rapid and cannot be the rate-determining step.

Looking at the two experiments together, it is interesting to know why the reaction is slow when the solid slag is added to the metal. If the fusion of the slag is a slow

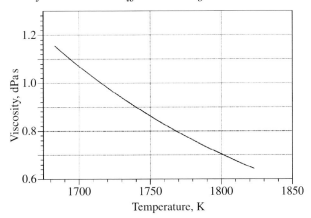

Figure 4. Variation of the viscosity of CaO–FeO–SiO$_2$ slag as a function of temperature in the range 1420–1550 °C.

process, then the amount of liquid slag available for dephosphorization would be limited, thus a slower dephosphorization rate. However, this possibility seems unlikely as the experimental temperature of 1550 °C is greatly above the liquidus temperature of the initial slag according to the literature (Kowalski *et al.* 1995). Another important factor may be the high viscosity of the slag as the molten slag heats up. The effect of temperature on the viscosity of the present slag in the temperature range 1410–1550 °C, assuming all iron oxide present is in the form of FeO, as calculated by the viscosity model available in the Division of Theoretical Metallurgy (Seetharaman & Du 1997) is shown in figure 4. It is seen that the viscosity decrease is substantial as the temperature rises. If this affects the rate of the reaction, it would lead to the conclusion that the dephosphorization reaction is likely to be controlled by the mass transfer in the slag phase.

4. Study of As and Sb transfer from molten copper to liquid Na$_2$CO$_3$

The abatement of As and Sb from copper is followed by changes in the oxygen content of the Cu melt. Oxygen is a strong surface-active element in liquid copper, as shown by Monma & Suto (1961). Thus, de-arsenification and de-antimonization can be expected to be accompanied by changes in the interfacial tension.

(a) Experimental procedure

Copper metal containing As or Sb was premelted in the sessile drop unit in Al$_2$O$_3$ crucibles at 1200 °C. The atmosphere was purified argon. Solid Na$_2$CO$_3$ was added from the top and the changes in the drop shape were observed. Some experiments were performed in CO$_2$ atmosphere instead of Ar.

(b) Results and discussion

Figure 5a shows the initial shape of the copper drop containing 0.2 wt% O̲ and 0.56 wt% As in Ar atmosphere before the addition of the slag. After 23 s of Na$_2$CO$_3$ addition, a momentary decrease in the interfacial tension was observed, as seen in figure 5b. However, this momentary decrease was not reproducible and needs further experimentation. The shape of the drop was restored to its original shape very quickly at 72 s and remained so even after 372 s (figure 5c–f). In a parallel chemical kinetics experiment carried out at Imperial College (Mangwiru & Jeffes, personal

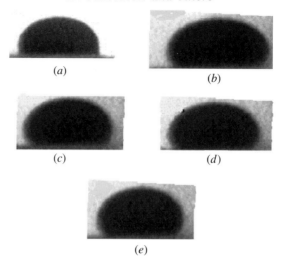

(a)

(b)

(c)   *(d)*

(e)

Figure 5. X-ray images of a copper drop with 0.2 wt% O and 0.56 wt% As in an Na_2CO_3 melt kept in an Al_2O_3 crucible at 1200 °C at various time intervals in an Ar atmosphere (solid Na_2CO_3 was added to liquid metal).

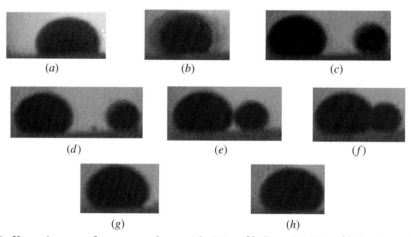

(a)   *(b)*   *(c)*

(d)   *(e)*   *(f)*

(g)   *(h)*

Figure 6. X-ray images of a copper drop with 0.2 wt% O and 0.56 wt% As in an Na_2CO_3 melt kept in an Al_2O_3 crucible at 1200 °C at various time intervals in a CO_2 atmosphere (solid Na_2CO_3 was added to liquid slag).

communication) by sampling and analysis, it was found that at the initial oxygen and arsenic levels in the copper metal, the arsenic removal was over within the first 100 s, which explains the difficulty in reproducing the rapid changes in the drop shape. In one of the experiments where CO_2 atmosphere was used, the splitting of the drop was noticed (figures 6a–h), which indicates the excessive mass-transfer reaction. The observations in the case of de-antimonisation were somewhat similar.

5. Summary

Dynamic interfacial tension studies are found to be very useful tools in following the metal–slag reactions and correlating the changes in the interfacial tensions and the reaction mechanisms. It was found that, with a sensitive sessile-drop unit, it should be possible to record the reaction sequences even during very short intervals of time. This provides an entirely new perspective of the micromechanisms of metal–slag reactions. It is hoped that, with parallel work at different laboratories in the world to ensure the reliability of the technique and the accuracy of the results, the sessile-drop technique will enable a better understanding of the kinetics and mass transfer phenomena in metal–slag reactions.

The authors thank Professor J. H. E. Jeffes, for providing us with the information. We also thank Professor J. Carlsson, President of the Royal Institute of Technology, for the partial financial support received for this work.

References

Gaye, H., Lucas, L. D., Olette, M. & Riboud, P. V. 1984 *Can. Metall. Q.* **23**, 179–191.

Kowalski, M., Spencer, P. J. & Neuschütz, D. 1995 *Slag-atlas* (ed. V. D. Eisenhüttenleute), 2nd edn, pp. 154–155. Germany: Stahleisen.

Kozakevitch, P. 1957 *Symp. on Physical Chemistry of Steelmaking*, p. 134. MIT Press.

Monma, K. & Suto, H. 1961 *Trans. Jap. Inst. Metals* **2**, 148.

Nasu, M., Mills, K. C., Monaghan, B., Jakobsson, A. & Seetharaman, S. 1997 Presented at the *5th Int. Conf. on Molten Slags, Fluxes and Salts '97, Sydney, Australia, 5–8/1/1997*.

Seetharaman, S. & Du, S. 1997 *ISIJ Int.* **37**, 109–118.

Studies on bubble films of molten slags

By Colin Nexhip, Shouyi Sun and Sharif Jahanshahi

*G. K. Williams Cooperative Research Centre for Extractive Metallurgy,
CSIRO-Division of Minerals, Box 312, Clayton South Victoria 3169, Australia*

Bubble films of CaO–SiO_2–Al_2O_3 slags were withdrawn on wire frames, and their draining rates measured at temperatures up to 1723 K using a developed gravimetric technique. The influence of additions of Na_2O and P_2O_5, as surfactants, on the draining rate and stability of slag films is investigated. Characteristic draining behaviour (interfacial flow) and the thickness of the films were investigated using a laser interferometry and laser absorption/transmission method. Evidence of localized thinning regions within the slag films, and measurements of the actual thickness of the films before rupture will be presented.

Keywords: foams; Plateau border; film thickness; interferometry; laser absorption

1. Introduction

In a well drained foam of soap solution, the plane lamellae intersect to form 'Plateau borders' (Plateau 1861). To some extent, this is also expected for a slag foam. Due to the curvature of the surface near the Plateau border, the pressure within these borders (P'') is much lower relative to the flatter central lamellae (films), where the pressure is essentially atmospheric (P'). The resulting pressure differential ($\Delta P = P' - P''$) creates a 'suction' force for the liquid to drain from the films into the Plateau borders under capillarity, and upon reaching the borders the liquid then proceeds to drain under the influence of gravity. For very thin films (how thin they are depends on the chemistry of the solution), a 'disjoining pressure' Π, originally suggested by Derjaguin & Titievskaya (1953), could become significant to counter the suction due to the curvature of the surface. The net capillary force is therefore given by $\Delta P = 2\gamma/r - \Pi$, where γ is the liquid surface tension and r is the radius of curvature of the liquid surface at the Plateau border. While the film draining characteristics and draining are extensively studied for aqueous systems (Mysels *et al.* 1959), our understanding of the characteristics of molten slag films is far from satisfactory. This paper presents some results of the study in determining the draining rate, the thickness, and thickness distribution of molten slag films spun on platinum wire frames. The slag used was CaO–SiO_2–Al_2O_3, and P_2O_5 or Na_2O was added to show the effect of surfactant on the draining of slag films.

2. Experiment

Measurement of the draining rate of slag films was achieved using a previously described gravimetric technique (Nexhip *et al.* 1997), which involved suspension of a rectangular wire frame inside a furnace hot-zone from an electronic balance (situated above). The molten slag sample was contained in a Pt-30% Rh crucible, which could

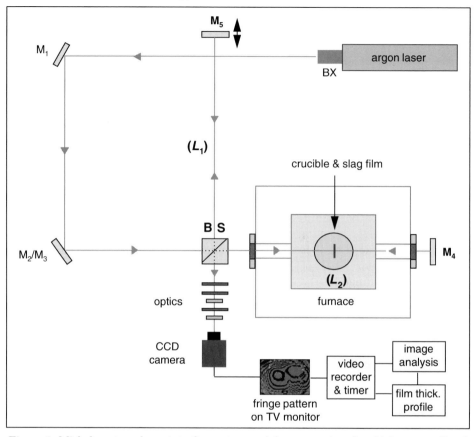

Figure 1. Michelson-type laser interferometer used for measuring the thickness profile of thinning slag bubble films.

be vertically raised or lowered using a motion actuator. Slag bubble films on the wire frames could then be formed by immersing the frame into the molten slag, and then partly withdrawing a film by lowering the crucible. As the films drained, the change in the gross mass was recorded as a function of the drain time (t_{d}) until rupture occurred. The film draining rate could then be calculated from the gravimetric data. For instance, for a given slag temperature, a linear regression of the drainage data (change in mass) over a given time interval was performed, yielding the draining rate of the slag film V_{f} $(\mathrm{g\,s^{-1}})$. The drain time intervals used in the rate calculations were usually about 5 s into a draining experiment, i.e. well after the withdrawal stage where the initial recorded mass increased. For a given slag composition, up to six consecutive draining experiments were usually recorded at a given temperature and the overall rates subsequently calculated from the average of these six runs. In order to investigate the film draining behaviour in the absence of surface tension phenomena, the viscous drainage of slag films from a thin vertical plate (Jeffreys 1930) was also investigated gravimetrically; this involved withdrawing films of identical area to that of the free films.

Interferometry is a technique that has been used extensively in studies of thinning bubble films in aqueous systems (Bikerman 1973). The experimental set-up of a Michelson-type laser interferometer (Meyer-Arendt 1972), shown in figure 1, consisted of an argon-ion laser source (Spectraphysics model 2017, operating at 300 mW and $\lambda = 488$ nm), a BK7 beamsplitter cube (BS), and two mirrors M_4 and M_5 (reflectivity greater than 99%). The cube split the already expanded laser beam into two orthogonal beams, each with an approximate beam diameter of 1.0 cm. The 'test' beam of physical pathlength (L_2) was transmitted through optical-quality silica windows, through the furnace hot-zone (and suspended wire frame), toward mirror M_4. The 'reference' beam of physical pathlength (L_1) was reflected orthogonally toward mirror M_5, which was mounted on a horizontal translation stage. The two mirrors M_4 and M_5 then retro-reflected the beams, which subsequently recombined at BS where interference between the two wavefronts could be observed (when coincident) through an optics train and monochromatic CCD video camera (Pulnix model TM-6CN). After equalizing the pathlengths of L_1 and L_2 (by translating M_5), the 'zero-order' reference fringe pattern was obtained. From here on, any change in the physical pathlength between the two interfering beams, introduced by the presence of a partly withdrawn thinning slag film, will cause a change in the relative phase of the two beams when they recombine at the beamsplitter cube; the change in phase being directly proportional to the thickness change within the film.

The thickness change ($\Delta\delta$) at a given point within a given time interval could then be calculated thus (Schlesinger *et al.* 1986):

$$\Delta\delta = (m\lambda)/2(\eta_f - 1), \tag{2.1}$$

where m is an integer and represents the number of fringes counted sweeping across the point in a given time from the TV monitor during film thinning, λ is the wavelength of laser light (488 nm) and η_f is the refractive index of the liquid slag. The factor 2 in the denominator arises due to the test beam traversing the slag film twice, and the factor 1 arises from the refractive index of air taken to be unity. No data for η_f are available for the liquid slag used in this study. A value of 1.57 (Shelby 1985) for a glass of similar composition was used instead. The thickness change during the major part of the film life could then be calculated by dividing $\Delta\delta$ by the total drain time (Δt), over which the fringes were counted in real time, before rupture.

A laser absorption/transmission method (figure 2) was also used to measure the thickness of the films using the Beer–Lambert relationship (Takahashi & Shibata 1979):

$$I = I_0 e^{-\alpha\delta}, \tag{2.2}$$

where δ is the film thickness, I is the intensity of a He–Ne laser beam transmitting a thinning slag film (at $\lambda = 633$ nm), I_0 is the reference intensity of the laser itself (obtained with no film present in the beam path) and α is the absorption coefficient of the slag (m^{-1}) obtained from the measurement data of Susa *et al.* (1992).

3. Results

The influence of small additions of surface-active oxides such as acidic P_2O_5 and basic Na_2O on the draining rate (and hence stability) of 'free' films of CaO–SiO_2–Al_2O_3 slags at 1623 K, can be seen in figure 3 where the rates (V_f) were calculated over the drain time interval of 5–15 s. The draining rate of both P_2O_5 and Na_2O containing (CaO/SiO_2) ~ 0.70, 17 wt% Al_2O_3 slag films, decreased with increasing (dilute) concentration of surfactant, suggesting a decrease in the slag surface tension is likely to be responsible. The curve for the P_2O_5 containing slag films appears to 'bottom-out' at the very dilute surfactant concentration of 0.2 wt%, the rate

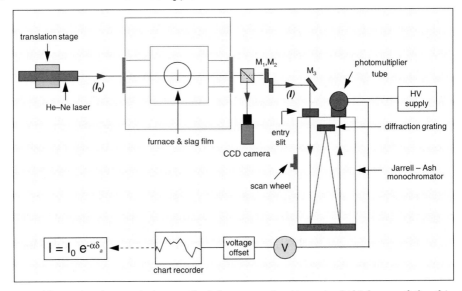

Figure 2. Absorption/transmission method for measuring the actual thickness of the thinning slag bubble films, before their rupture.

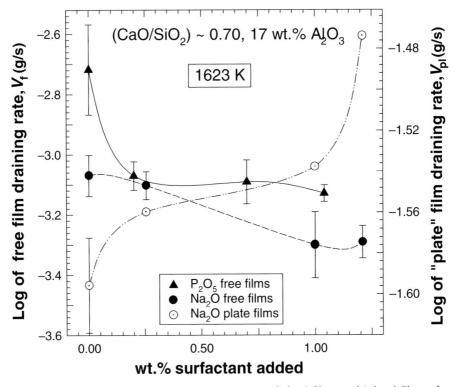

Figure 3. Calculated gravimetric draining rates of 'free' films and 'plate' films of $(CaO/SiO_2) \sim 0.70$ slags, as a function of surfactant concentration, at 1623 K.

decreasing by a factor of about three. This implies that the films may readily become saturated at very dilute concentrations of P_2O_5 in silica-rich slags. The decrease in the draining rate of Na_2O-containing films was not as dramatic, the rate decreasing by a factor of about 1.5 at 1.0 wt% Na_2O; however, the overall draining rates of the soda-containing films were lower than those for the phosphorus containing films. Presently, it is uncertain as to why the draining rates for the free films in which phosphorus and soda were absent are so different. For instance, the rate for the 0.0 wt% Na_2O films are much slower than the 0.0 wt% P_2O_5 films, despite their respective basicities being nearly identical at $(CaO/SiO_2) = 0.71$ and 0.70. CaO–SiO_2–Al_2O_3 slag films draining from a thin (vertical) platinum plate showed the opposite trend to that observed for the free films, i.e. the plate film draining rate (V_{pl}) increased with increasing concentration of Na_2O (figure 3). The gravimetric draining curve characteristics of the plate films (not shown) were also much smoother than those of the free films, suggesting they drain more uniformly in the absence of capillary effects.

Typical laser interference fringe patterns obtained for a thinning 'free' film of a $(CaO/SiO_2) = 0.60$, 15 wt% Al_2O_3 slag at 1623 K can be seen in figure 4. What is immediately apparent is that the film thickness profile is non-uniform, and a localized thinning spot within the slag bubble lamella is seen to grow radially, eventually forming two thinning regions. From the fringe patterns obtained, the thickness of the thicker sections appear to be a few μm greater than that of the thinner sections. From equation (2.1), the calculated thickness change during the major part of the film life was about 8 μm, suggesting a thinning rate of $\Delta\delta/\Delta t \sim 0.3$–$0.4\ \mu m\ s^{-1}$.

The results for the laser absorption/transmission experiments in figure 5 show that the measured thickness of thinning slag films decreases in the latter stage from several μm down to $\delta \sim 0.1$–$0.4\ \mu m$ before rupture. The figure shows the thickness at the 'centre' and 'edge' of a film to be different initially; however, during the well-drained stage the curves begin to coincide, indicating that the thickness profile across the films become very uniform. This suggests that thin (well-drained) slag films are likely to become plane-parallel before rupture, a characteristic observed within aqueous films around $\delta < 0.1\ \mu m$ (Platikanov 1964). Experiments on $(CaO/SiO_2) = 0.60$, 15 wt% Al_2O_3, 9.6 wt% Fe_2O_3 slags at 1573 K are shown in figure 6, again suggesting that the presence of dilute concentrations of P_2O_5 in acidic (silica rich) CaO–SiO_2–Al_2O_3 slags can increase the stability of bubble films. For instance, the slag films containing 0.8 wt% P_2O_5 ruptured at a critical thickness of about 0.38 μm (after a film life of 39 s), whereas those containing 1.3 wt% P_2O_5 generally ruptured at a lower critical thickness of about 0.15 μm (film life of 47 s).

4. Discussion

P_2O_5 has been shown to be surface active in CaO–SiO_2 slags at a $(CaO/SiO_2) = 0.69$ at temperatures of 1823–2050 K (Cooper & Kitchener 1959). The effect of Al_2O_3 is not clearly known but is not expected to dramatically change the surface activity of P_2O_5. Na_2O has also been shown to be surface active in CaO–SiO_2 slags (Swisher & McCabe 1964) and CaO–SiO_2–Al_2O_3 slags (Takayanagi *et al.* 1976), at $(CaO/SiO_2) \sim 0.8$. The influence of P_2O_5 on the draining rate of free films of $(CaO/SiO_2) = 0.70$, 17 wt% Al_2O_3 slags was analogous to that usually observed in aqueous systems, i.e. the rate decreased with decreasing surface tension, which is likely to be due to a decrease in the Plateau border suction force operating within the films. The observed increase in stability through addition of P_2O_5 to acidic CaO–SiO_2–Al_2O_3 slags could be due to the 'cosorption' of P_2O_5 in a silica rich surface environment (Cooper & McCabe 1961).

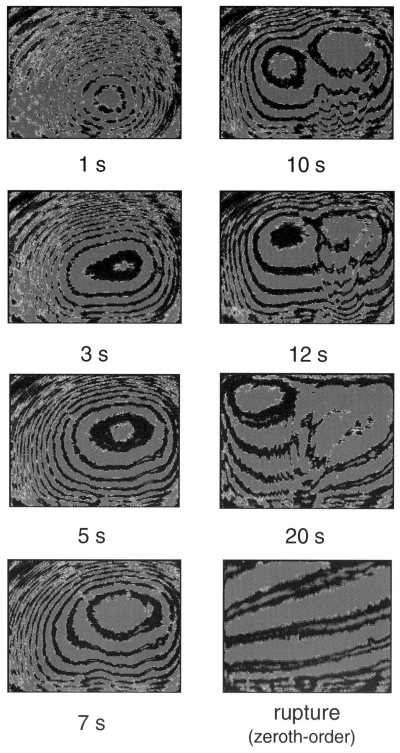

1 s

10 s

3 s

12 s

5 s

20 s

7 s

rupture
(zeroth-order)

Figure 4. Laser interference fringe patterns for a thinning $(CaO/SiO_2) = 0.60$, 15 wt% Al_2O_3 free film at 1623 K, showing a non-uniform thickness profile and localized thinning behaviour.

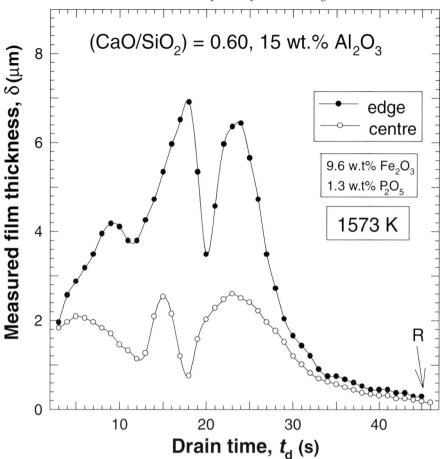

Figure 5. Laser absorption/transmission measurements of the thickness of free films of $(CaO/SiO_2) = 0.60$, 15 wt% Al_2O_3, 9.6 wt% Fe_2O_3 slags, showing the film profile becomes uniform at the well-drained stages, before rupture at a thickness less than 1 μm.

A high surface viscosity has been suggested to be important in increasing the stability of slag foams (Hara *et al.* 1990), as it retards liquid drainage from the lamellae. The results obtained for the addition of acidic P_2O_5 to CaO–SiO_2–Al_2O_3 slags agree with this statement, i.e. an increase in P_2O_5 decreased the film draining rate, which may have been due to increasing the surface viscosity of the bubble films (by polymerizing the surface). However, as already mentioned, the influence of small additions of Na_2O (as a basic surface-active oxide) to CaO–SiO_2–Al_2O_3 slags, was actually found to decrease the film draining rate. Thus the experimental results do not support the suggestion that surface viscosity significantly influences the draining rate of foam films, as the addition of Na_2O would be expected to depolymerize the surface layer and hence increase the film draining rate.

According to the present study, surface-tension lowering is likely to be the most important mechanism stabilizing the bubble films in a slag foam, as the draining rate decreased markedly when the surface tension was decreased with the addition of both acidic and basic surface-active oxides. Rather than the absolute value of the surface tension being important, the rate of change of surface-tension depression per unit concentration change of surfactant ($d\gamma/dC$) has been acknowledged to be a more

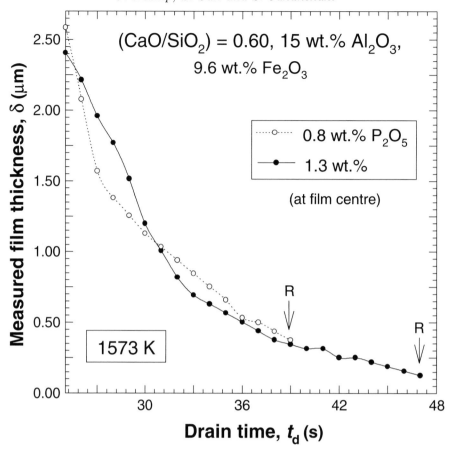

Figure 6. Influence of P_2O_5 addition on the extent of film thinning of $(CaO/SiO_2) = 0.60$, 15 wt% Al_2O_3, 9.6 wt% Fe_2O_3 slags at 1573 K. An increase in P_2O_5 concentration results in a decrease in the critical thickness and an increase in the lifetime/stability of the slag bubble films.

important criterion when determining the stability of slag foams. For instance, a high value of $d\gamma/dC$ implies the slag films to have an effective 'elasticity mechanism' which can enhance foam stability. In the CaO–SiO_2 system, the values calculated for P_2O_5 or Na_2O in slags of $(CaO/SiO_2) \sim 0.7$ are $d\gamma/dC \sim 11$ and 10 mN m^{-1}/mol%, respectively (Cooper & Kitchener 1959; King 1964). These values are quite high when compared to other 'non-foaming' systems such as $30Na_2O$-$70SiO_2$, with a $d\gamma/dC \sim 0.7$ mN m^{-1}/mol% (Hara *et al.* 1990).

The results from the laser interferometry experiments in figure 4 showed that the slag bubble films can in fact thin non-uniformly. The regions where thinning spots were observed $(\Delta\delta/\Delta t \sim 0.4 \, \mu\text{m s}^{-1})$, are likely to be influenced by Plateau border suction forces, as the local radii of film curvature is likely to change. Image processing using a fast Fourier transform (FFT) method, has been used by the present authors to analyse the spacial frequencies of the fringe patterns (Nexhip *et al.* 1997*b*). The results showed that the frequencies of the fringe patterns at the well-drained stages of the films, approached that of the 'zero-order' fringe pattern (obtained with no slag film present). As with the slag film profile previously suggested in figure 5 (using laser absorption/transmission), the results from the FFT method on interference fringe patterns also suggest the thickness across the films becomes uniform/plane-parallel before their rupture.

5. Conclusions

Studies on 'free' films of CaO–SiO_2–Al_2O_3 slags at $(CaO/SiO_2) \sim 0.7$, have revealed that the presence of strong surfactants such as acidic P_2O_5 and basic Na_2O markedly decrease the draining rate of the bubble films. Surface-tension lowering ability (or effective elasticity, $d\gamma/dC$) is likely to be the most important mechanism influencing the drainage of slag bubble films, i.e. a lower surface tension will result in a decrease in the Plateau border suction force. This conclusion was supported by the fact that the addition of Na_2O to the aforementioned slag actually increased the draining rate of 'plate' films, in which capillary forces are insignificant. The surfaces of free films of a $(CaO/SiO_2) = 0.60$, 15 wt% Al_2O_3 slag were also shown to be non-uniform in thicknesses, with localized thinning occurring within the lamella. At the well-drained stage, slag bubble films are likely to become plane-parallel, with rupture occurring at thicknesses around 0.1–0.4 µm.

Financial support for this work was provided by the Australian Mineral Industries Research Association (AMIRA) and Australian Government Cooperative Research Centre Program, through the G. K. Williams CRC for Extractive Metallurgy, a joint venture between the CSIRO (Division of Minerals) and the University of Melbourne (Department of Chemical Engineering).

References

Bikerman, J. J. 1973 *Foams*, pp. 65–97, 149–158. New York: Springer.

Cooper, C. F. & Kitchener, J. A. 1959 *JISI* **193**, 48–55.

Cooper, C. F. & McCabe, C. L. 1961 Physical chemistry of process metallurgy. *Met. Soc. Conf.* **7**, pt 1, 117–131.

Derjaguin, B. V. & Titievskaya, A. S. 1953 *Kolloid Zh.* **15**, 416.

Hara, S., Kitamura, M. & Ogino, K. 1990 *ISIJ Int.* **30**, 714–721.

Jeffreys, H. 1930 *Proc. Camb. Phil. Soc.* **26**, 204.

King, T. B. 1964 *Trans. Met. Soc. AIME* **230**, 1671.

Meyer-Arendt, J. R. 1972 *Introduction to classical and modern optics.* Englewood Cliffs, NJ: Prentice-Hall.

Mysels, K. J., Shinoda, K. & Frankel, S. 1959 *Soap films, studies of their thinning.* London: Permagon.

Nexhip, C. 1997 Fundamentals of foaming in molten slag systems. Ph.D. thesis, University of Melbourne.

Nexhip, C., Sun, S. & Jahanshahi, S. 1997 Drainage of molten CaO–SiO_2–Al_2O_3 slag films. *Proc. 5th Int. Conf. on Molten Slags, Fluxes and Salts, Sydney*, p. 297. Warrendale, PA: Iron and Steel Society.

Plateau, J. 1861 *Mem. Acad. R. Sci. Belg.* **33**, 5th and 6th series.

Platikanov, D. 1964 *J. Phys. Chem.* **68**, 3619.

Schlesinger, M., De Kee, D. & Godo, M. N. 1986 *Rev. Sci. Instrum.* **57**, 2535–2537.

Shelby, J. E. 1985 *J. Am. Ceram. Soc.* **68**, 155–158.

Susa, M., Futao, L. & Nagata, K. 1992 *Proc. 4th Int. Conf. on Molten Slags and Fluxes, Sendai*, pp. 22–27. ISIJ, 1-9-4, Otemachi Chiyoda-ku, Tokyo, Japan.

Swisher, J. H. & McCabe, C. L. 1964 *Trans. Met. Soc. AIME.* **230**, 1669–1675.

Takahashi, S. & Shibata, S. 1979 *J. Non-Crystall. Solids* **30**, 339.

Takayanagi, T., Kato, M. & Minowa, S. 1976 *J. Jap. Foundry Soc.* **48**, 779.

Discussion

S. K. WILSON (*Department of Mathematics, University of Strathclyde, Glasgow, UK*). Could Professor Nexhip give us an indication of the typical size of surfactant gradients in the experiments, i.e. whether or not the role of the surfactant is primarily to change the mean value of surface tension or to create surface-tension gradients? If surface-tension gradients are significant then the author might be interested in contacting Dr J. R. Ockendon's group at the Oxford Centre for Industrial Applied Mathematics (Oxford University), who are currently undertaking a theoretical investigation of the effects of surface-tension gradients in foam drainage.

C. NEXHIP. Dr Wilson has certainly identified a very interesting issue in the capillary-driven flow in a liquid film. Both the lowering of the surface tension by the addition of a surfactant and the non-uniform distribution of the surfactant on the surface could have implications on the draining of liquid films. However, to our best knowledge, there have been no direct measurements of the 'surface' composition of a molten slag. Almost all information regarding the surface composition of a molten slag is derived from surface-tension data together with the Gibbs adsorption isotherm. In other words, all these refer to the equilibrium state. Under transient or dynamic conditions, spatial variation of the surface tension may be taken to imply that a surfactant gradient exists on the surface. Unfortunately, no conclusive experimental knowledge is available regarding this gradient.

   In summary, the surface excess of surfactants in a molten slag is usually inferred from equilibrium measurements. No firm indication can be given for the surfactant gradient. For that reason, other approaches to the investigation in surfactant gradients (such as the suggested work by Dr J. R. Ockendon's group) could be interesting.

M. McLEAN (*Department of Materials, Imperial College, London, UK*). The big difference in the draining rates of two nominally identical slags with *no* added surfactant requires some comment (see figure 3). Does this indicate that the results are particularly sensitive to the slag composition (apart from surfactant) or that the measurements are due to a high level of uncertainty?

C. NEXHIP. The P_2O_5 and Na_2O-free slags had the same target composition. However, as shown in figure 3, the draining rate for the 'surfactant free' slag from the Na_2O series was noticeably lower than that from the P_2O_5 series of experiments. It is currently uncertain as to why the two draining rates were so different. Two factors could have contributed to it. Firstly, the samples used were from two batches. All samples were prepared from a premelted master slag of 18% Al_2O_3 with $CaO/SiO_2 = 0.93$, by adding a specified amount of SiO_2 to get the desired CaO/SiO_2 ratio. Unlike the sample for the P_2O_5 series, which was prepared from a fresh batch of the master slag, the one for the Na_2O experiments was from a batch that had been used earlier in a series of 'surfactant-free' draining experiments. Secondly, the Na_2O series of experiments were carried out after the P_2O_5 experiments, and there was some evidence of P_2O_5 loss in the earlier experiments. Thus any P_2O_5 deposited on the furnace interior could have become a surface contaminant in the Na_2O-free containing slags (used subsequently).

Modelling of the erosion of refractories by Marangoni flows

By G. Tsotridis[1] and E. D. Hondros[2]

[1]*Institute for Advanced Materials, Joint Research Centre, E.U., PO Box 2,*
1755 ZG Petten, The Netherlands
[2]*Department of Materials, Imperial College of Science, Technology and Medicine,*
London SW7 2BZ, UK

This study is concerned with the industrially important problem of the erosion and degradation of ceramic crucibles containing liquid metals or glasses. A model has been developed which can describe the patterns of erosion in terms of the interplay between Marangoni and buoyancy forces forces in the melt and the disruption of the concentration gradients of dissolved species from the containment wall. The model predicts significant erosion effects which are highly system dependent and thus the elaboration of the model and its useful application depends critically on the availability of capillarity and thermochemical data relevant to the complex systems under study.

Keywords: flux-line erosion; Marangoni flows; capillarity;
Navier–Stokes equation; refractory crucibles

1. Introduction

The phenomenon often referred to as 'flux-line erosion' has long been recognized in the metal and glass production industries as a problem for ceramic containment vessels. This consists essentially of the scouring erosion at high temperatures of the inner walls of the refractory crucible, generally where the interface between the liquid metal and slag, or the glass and its layer of flux, meets the wall. A variety of erosion profiles has been encountered—for example, 'waist' formation, 'necking', 'cut' formation—these shapes depending on the temperature and in particular, the chemistry of the solid and molten phases. It is a highly destructive process, leading to the degradation of the vessel and its costly replacement.

We have developed a mathematical treatment for this process based on the model of a partly immersed solid in a reactive liquid which undergoes slow dissolution, giving rise to a concentration gradient of relevant solute species away from the wall (Tsotridis *et al.* 1992). The erosion pattern is governed by the interference and disruption of the concentration gradients by convective streaming arising from buoyancy and surface tension forces. Furthermore, the model assumes that the material of the containment vessel is homogeneous down to molecular levels, that is, effects due to microstructure, inclusions and grain boundaries are discounted. The dissolved anion or cation species from, for example, an oxide ceramic wall may adsorb on liquid phase interfaces, giving rise to localized interfacial tension gradients which actuate the Marangoni flows, the direction of flow depending on whether the solutes are positively or negatively adsorbed. The model assumes these currents can sweep away the dissolved wall species, thus exposing or blocking fresh wall material to the dissolution process.

In addition to the above capillary driven flow mechanism, there is an interplay with a more conventional convective flow mechanism. This arises from the fact that dissolved species in the melt adjacent to the vertical wall will change the physical properties of the melt, in particular the localized melt density, thereby leading to concentration buoyancy flows. The relative strength and interactions between these basic flow types will determine the rate of wall erosion and the shape of the erosion profile.

The model has been formulated by coupling the Navier–Stokes equation with the diffusion equation. The velocity field is determined by the equations of motion and conservation of mass, whereas the concentration distribution is determined by the convective diffusion equation. These equations are mutually dependent since the surface tension is a function of concentration. The concentration distribution depends on the velocity distribution, the driving forces for the flow deriving from adsorption-induced surface tension and buoyancy forces due to density gradients.

In this study, we have developed a two-dimensional transient flow computer program which solves simultaneously the equations of motion and diffusion. This allows the delineation of the flow patterns inside the melt as well as the concentration distributions. Positive Marangoni values tend to create erosion patterns in the immediate vicinity of the free surface. The study shows that even very low values of surface tension gradients produce remarkable effects on the patterns of erosion which are similar to those observed in practice for certain system combinations.

The surface or interfacial tensions resulting from diffusive concentration gradients are, of course, highly system sensitive and, faced with a dearth of relevant capillarity and thermochemical information for complex systems, the use of the present model in a predictive sense is, at this stage, disappointingly limited. First, there is required an assessment of the validity of the present model, which is in fact being carried out currently on a simple system with few components and for which the surface tension and all requisite thermochemical information is being separately measured. Following this, and with a programme of measurement of physical and thermochemical data on complex practical systems, we believe that it will eventually be possible to predict those sensitive surface active species which must be controlled in a given system in order to minimize erosion degradation, and hence extend the life of industrial refractories.

References

Tsotridis, G., Rother, H. & Hondros, E. D. 1992 Marangoni flows and the erosion of ceramic crucibles. *Naturwissenschaften* **79**, 314–317.

Marangoni flows and corrosion
of refractory walls

By Kusuhiro Mukai

Department of Materials Science and Engineering, Kyushu Institute of Technology,
1-1, Sensui-cho, Tobata-ku, Kitakyushu 804, Japan

Local corrosion of a refractory at slag–gas and slag–metal interfaces is caused by the active motion of slag film formed by the wettability between the refractory and slag. In the systems of $SiO_2(s)$–$(PbO$–$SiO_2)$ slag and $SiO_2(s)$–$Pb(l)$–$(PbO$–$SiO_2)$ slag, where components from the refractory dissolve into the slag increasing the interfacial tension of the slag or slag–metal, continuous washing of the refractory wall with thin fresh slag film supplied from the bulk slag phase by the Marangoni effect causes the local corrosion of the refractory above the slag level or below the metal level. When the dissolved component from the refractory reduces interfacial tension, for example, in the system of $SiO_2(s)$–$(Fe_tO$–$SiO_2)$ slag, the local corrosion is induced by the active slag-film motion with alternative formation and disappearance in cyclic mode due to the Marangoni effect, and changes in the form of slag film due to the variation of interfacial tension and density of the slag film. Local corrosion of oxide refractories for practical use, containing trough materials, at slag–gas and slag–metal interfaces is classified into the above two types. The Maragoni flow of the slag film is also considered to play an important role in the local corrosion of oxide–graphite refractories such as Al_2O_3–C and MgO–C at the slag–metal interface during the stage of the oxide phase corrosion in the refractory.

Keywords: local corrosion; refractory; iron and steel making; slag surface;
slag–metal interface; Marangoni flow

1. Introduction

It is well known that refractories composed of oxides, oxide–graphites and oxide–graphite–carbides are corroded locally at the slag–gas (slag surface) or slag–metal interface in glass technology and during iron and steel making processes. The local corrosion of refractories is a serious problem for these industries because it limits the life of the refractories. Several ideas had been proposed on the mechanism of the local corrosion. Jebsen-Marwedel (1956) proposed that the local corrosion of solid oxide by molten glass is caused by interfacial turbulence of the molten glass induced by the Marangoni effect in the vicinity of the interface. Vago & Smith (1965) explain that a reactive vapour phase from a liquid glass plays an important role in the local corrosion of solid oxide, with the form of a neck above glass level. Caley *et al.* (1981) state that oxygen gas in the atmosphere is the main cause for the local corrosion of solid oxide by molten PbO–SiO₂ slag. Brückner (1967), Sendt (1965) and Schulte (1977) have estimated that local corrosion of the solid oxide at the slag–metal interface is also caused by an interfacial turbulence of molten glass induced by the Marangoni effect in the vicinity of the interface. Iguchi *et al.* (1979)

201

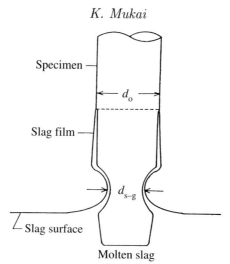

Figure 1. Typical local corrosion profile of the $SiO_2(s)$–$(PbO–SiO_2)$ slag system (Mukai *et al.* 1983).

explained the local corrosion of solid alumina at the slag–metal interface based on the model of electrochemical reaction. Since 1983, authors have been investigating the local corrosion of refractories at the interfaces of slag–gas and slag–metal by combining optical or X-ray radiographic technique-aided direct observation of the phenomenon, which occurs in the local corrosion zone with a conventional immersion test in the following systems: (1) solid silica $(SiO_2(s))$–$(PbO–SiO_2)$ slag; (2) $SiO_2(s)$–$Pb(l)$–$(PbO–SiO_2)$ slag; (3) $SiO_2(s)$–$(Fe_tO–SiO_2)$ slag; (4) $SiO_2(s)$–$(Na_2O–SiO_2)$ slag; (5) magnesia-chrome refractory–$(CaO–Al_2O_3–SiO_2)$ slag; (6) magnesia-chrome refractory–$Fe(l)$–$(CaO-Al_2O_3–SiO_2)$ slag; (7) blast furnace trough material–$(CaO–Al_2O_3–SiO_2)$ slag; (8) the trough material–Fe–C alloy(l)–$(CaO–Al_2O_3–SiO_2)$ slag; (9) Al_2O_3–C refractory–Fe-C alloy(l)–$(CaO–Al_2O_3–SiO_2)$ slag; and (10) MgO–C refractory–$Fe(l)$–$(CaO–Al_2O_3–SiO_2)$ slag.

2. Local corrosion of $SiO_2(s)$–slag and $SiO_2(s)$–Pb(l)-slag systems

(a) $SiO_2(s)$–$(PbO–SiO_2)$ *slag system (Mukai et al. 1983, 1984a, 1985, 1986a,b)*

Immediately after partially immersing the SiO_2 specimen in $PbO–SiO_2$ slag at 1073 K in an Ar atmosphere, slag begins to creep up the specimen surface to form slag film above the slag level. The active slag-film motion can be directly observed and recorded on cinefilm with the aid of the movement of very small bubbles in the slag film. The local corrosion of this system, as shown in figure 1, progresses in the area where the slag film moves actively. When platinum wire is tightly wound on the surface of the SiO_2 specimen at, and around, the slag level, the wire prevents the slag film moving and the local corrosion does not occur. Oxygen content in the atmosphere does not affect the local corrosion rate. The above experimental results indicate that neither a reactive vapour phase, nor oxygen in the atmosphere, is the main cause of the local corrosion of this system.

The slag film forms characteristic flow patterns, principally composed of wide zones of rising film and narrow zones of falling film, according to the contour of the specimen, as shown in figures 2 and 3. In the case of a cylindrical specimen, the position of the falling zone moves gradually on the surface of the specimen. However,

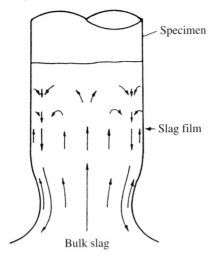

Figure 2. Typical flow pattern of the slag film on the 6 mm OD silica specimen dipped in PbO–SiO$_2$ slag (Mukai *et al.* 1985).

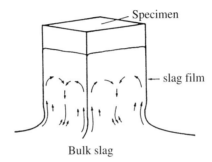

Figure 3. Typical flow pattern of the slag film of a 6 mm square prism specimen (Mukai *et al.* 1985).

the prism-shaped specimen always has one zone of rising film at, and around, each corner of the specimen and one narrow zone of falling film at each plane side of the specimen, as shown in figure 3.

Rising-zone films are several times thinner than falling-zone films. Change in SiO$_2$ content in rapidly solidified slag film is detected perpendicular to the specimen surface and also along the surface of the specimen in the vertical direction. Since contact time of the upper film with the specimen is longer than that for the lower film, the upper film has a higher SiO$_2$ content, due to the dissolution of SiO$_2$ from the specimen into the film. The difference in SiO$_2$ content causes a surface tension gradient in the vertical direction; for surface tension of PbO–SiO$_2$ slag increases with SiO$_2$ content (Hino *et al.* 1967). Thus the slag film is continuously pulled up by the surface tension gradient, i.e. the rising-zone of the film is generated by the Marangoni effect due to the concentration gradient. Since the film motion in the upper part of the specimen is slower than the lower part due to its high viscosity coefficient, the risen slag is accumulated in the upper part. The thickened film caused by the accumulation begins to fall down when its weight exceeds the surface tension gradient, i.e. a film falling zone is formed. Hydrodynamic analysis of the film motion, using the observed surface tension gradient along the surface of the film, results in good agreement between the calculated and observed film velocities. Velocity distribution

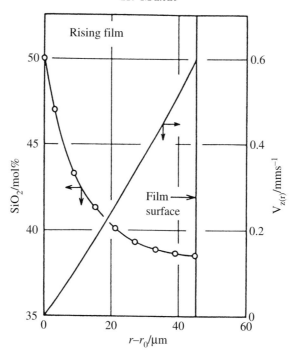

Figure 4. Distribution of velocity and SiO$_2$ content in the slag film for a 6 mm OD specimen dipped for 0.3 ks in a PbO–30 mol%SiO$_2$ slag at 1073 K (Mukai *et al.* 1985).

in the slag film obtained from the analysis shows that the slag film motion in the local corrosion zone is still active even in a thin slag film with a thickness of several tens of μm (figure 4). The film has a concentration gradient of SiO$_2$ perpendicular to the surface of the solid silica, which means that a diffusion layer is formed over the whole range of the slag film. The corrosion rate in the rising zone of the slag film is well-described by the rate equation derived from the conditions that the rate is controlled by mass transport of the dissolved component SiO$_2$ in the slag film under the Marangoni flow of the film. Therefore, the Marangoni flow in the slag film results in breakdown of the diffusion layer, which is the main cause of local corrosion. In other words, the local corrosion proceeds largely as a result of the wall-washing of solid silica with a fresh thin, rising slag film induced by the Marangoni effect. The corner of a prism specimen that is continuously washed always by the rising film, as shown in figure 3, is thus corroded much faster than the plane side, which leads to the formation of a round horizontal specimen cross section from an initial square-shaped cross section. On the other hand, a cylindrical specimen retains its initial round horizontal cross section in the local corrosion zone, because the rising zone of the slag film shifts its location with time on the surface of the specimen horizontally, resulting in an almost even corrosion rate at the same level in the local corrosion zone.

(b) *The* SiO$_2$(s)–(Fe$_t$O–SiO$_2$) *slag system (Yu & Mukai 1992; Yu 1995)*

Surface tension of PbO–SiO$_2$ slag is increased by the dissolved component SiO$_2$, as mentioned above, while SiO$_2$ reduces the surface tension of Fe$_t$O–SiO$_2$ slag (Yu 1995). Local corrosion and slag film motion in the local corrosion zone of this system can also be directly observed because the major part of the local corrosion zone

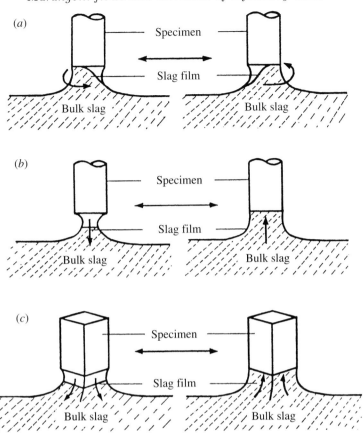

Figure 5. Slag film movements for rod and prism silica specimens dipped in a Fe_tO–SiO_2 slag (Yu & Mukai 1992): (*a*) rotational movement of slag film for the rod specimen; (*b*) up-and-down movement of slag film for the rod specimen; (*c*) up-and-down movement of slag film for the prism specimen.

occurs in the meniscus region of the slag (figure 5). The local corrosion zone of this system forms a steep groove and the vertical width of the local corrosion zone is narrower than that of the $SiO_2(s)$–$(PbO$–$SiO_2)$ slag system. The slag film of this system has two types of characteristic motion. One is a rotational motion around the specimen surface during the initial stage of the local corrosion of the cylindrical specimen. The other is an up-and-down motion of the whole slag film along the specimen surface during the developed stage of the local corrosion of the cylindrical specimen, and during the entire stage of the corrosion of the prism specimen, as shown in figure 5. When the slag film motion is observed at a fixed point on the surface of the specimen, the film motion shows the following cyclic mode from stage 1 to 4: the slag film thickness (1) increases; (2) reaches a maximum state; (3) decreases; and (4) arrives at a minimum state, which is shown in figure 6. The slag film is formed during stage 1 within the local corrosion zone and disappears during stage 3. Figure 6 indicates that the maximum film thickness increases with increasing immersion time. The period of the cyclic motion increases linearly with increasing immersion time, which is qualitatively found from the increase in the space between adjacent lines in figure 6 with increasing immersion time. The shape of the slag film surface (meniscus shape) formed in the local corrosion zone is approximately described by a Laplace

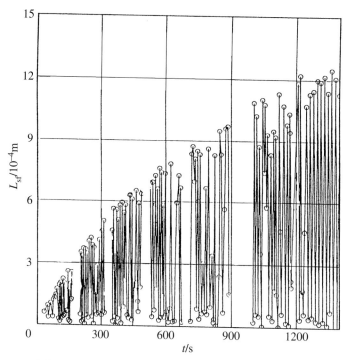

Figure 6. Change of slag film thickness in maximum local corrosion area, L_{sf}, with dipping time, t for a 75 mass%FeO–25 mass%SiO$_2$ slag with rod diameter 6 mm at 1623 K (Yu 1993).

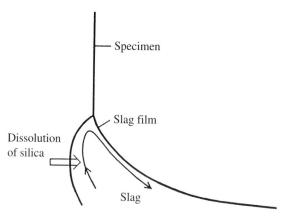

Figure 7. Marangoni convection of slag film in the local corrosion zone of a
SiO$_2$(s)–(FetO–SiO$_2$) slag system.

equation, which includes σ/ρ, contact angle θ between slag and solid silica and rod diameter D as its main variables. σ and ρ are the surface tension and density of the slag, respectively. When θ is nearly constant, which is experimentally observed in this system, the slag film height (distance from slag level to the upper end of the slag meniscus) increases with increasing σ/ρ and D. The SiO$_2$ content in rapidly solidified slag film at stages 1 and 2 is lower than at stages 3 and 4, which means that σ/ρ is higher at stages 1 and 2 than at stages 3 and 4. Therefore, during stages 1 and 2, slag creeps up on the surface of the specimen and forms slag film and then reaches a maximum height. During stages 3 and 4 the slag film comes down and disappears.

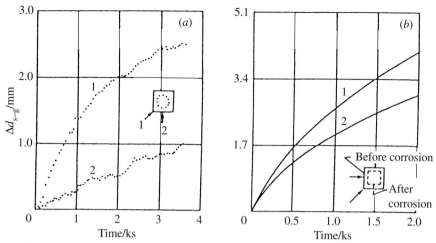

Figure 8. Comparison of the linear loss, Δd_{s-g}, at the planar part with that at the corner part of a prism specimen for: (a) $SiO_2(s)$–$(PbO$–$SiO_2)$ slag, $T = 1073$ K (Mukai *et al.* 1983); (b) $SiO_2(s)$–$(Fe_tO$–$SiO_2)$ slag $T = 1073$ K (Yu & Mukai 1992). $\Delta d_{s-g} = d_0 - d_{s-g}$ (see figure 1).

During stages 1–4, the SiO_2 content in the upper part of the slag film surface is higher than in the lower part of the surface. The SiO_2 content of the film decreases, reaches a minimum and then increases with distance from the slag film–specimen interface. The above change in SiO_2 content in the film indicates the continuous occurrence of the Marangoni flow in the film, as shown in figure 7, which accelerates the dissolution rate of the specimen into the slag film. The cyclic mode, with the formation and disappearance of the slag film, can be regarded as an analogous phenomenon to the surface renewal model proposed by Danckwerts (1951). The observed local corrosion rate of this system is reasonably explained by surface renewal theory. The corrosion rate at the corner of the prism specimen is $\sqrt{2}$ times as large as that at the plane. The cross-section of the prism specimen remains square during the entire corrosion process because the slag film has the same up-and-down motion at both the plane and corner sides, while in the $SiO_2(s)$–$(PbO$–$SiO_2)$ slag system, the horizontal cross section of the prism specimen changes its shape from square to round, as shown in figure 8.

(c) *The $SiO_2(s)$–(Na_2O) slag system (Mukai & Yu 1993, unpublished work)*

SiO_2 scarcely decreases the surface tension of Na_2O–SiO_2 slag (King 1964). When the SiO_2 specimen is partially immersed in this slag, a slag film is also formed above the slag level. However, neither film motion nor the local corrosion is detected experimentally in this system.

(d) *The $SiO_2(s)$–$Pb(l)$–$(PbO$–$SiO_2)$ slag system (Mukai et al. 1984b; 1989a)*

The Marangoni flow of the slag film in the local corrosion zone of the $SiO_2(s)$–$Pb(l)$–$(PbO$–$SiO_2)$ slag system is induced by the concentration gradient of SiO_2 in the film at the interface between liquid lead and the slag film in the vertical direction (figure 9). The slag film motion was observed directly through the wall of a transparent silica crucible. Figure 9 shows that the flow pattern is composed primarily of a downward flow during the initial stage of the corrosion and then an upward film flow increases its frequency and area with time. The active Marangoni flow of the slag film with a thickness of several tens of μm accelerates the dissolution

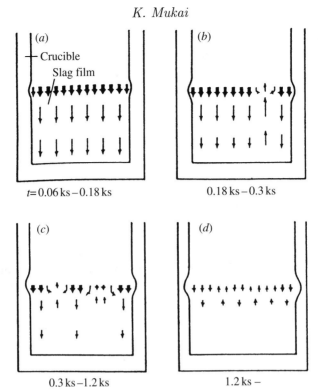

Figure 9. Schematic representation of the flow pattern of PbO–SiO$_2$ slag film formed between liquid lead and a silica crucible: t, elapsed time after the formation of the film (Mukai *et al.* 1989*a*).

rate of SiO$_2$ from the SiO$_2$ crucible into the film, resulting in local corrosion at the metal–slag interface. The manner in which local corrosion occurs in this system is essentially the same as that for the SiO$_2$(s)–(PbO–SiO$_2$) slag system when the phase of liquid lead is replaced with a gas phase and the present system is turned upside down.

3. Local corrosion of refractory for practical use: slag and the refractory–metal–slag systems

(*a*) *The magnesia–chrome refractory–(CaO–Al$_2$O$_3$–SiO$_2$) and the refractory–Fe(l)–slag system (Zainan et al. 1996; Mukai et al. 1995, unpublished work)*

The main corrosion modes of the magnesia–chrome refractory by the molten slag in the static condition are the local corrosion at interfaces of slag–gas and slag–metal, the pit-like corrosion at the bottom surface of the refractory specimen in the bulk slag phase and also in the upper part of the local corrosion zone. These corrosion modes were observed optically or by the aid of X-ray radiographic techniques.

The local corrosion rate at the slag surface increases with the iron oxide concentration of the slag and a decrease in the MgO and Al$_2$O$_3$ concentrations of the slag.

Both the local and pit-like corrosions are mainly caused by the Marangoni convection induced by the surface or interfacial tension difference along the interface of the slag film which forms on the refractory specimen of the local corrosion zone, or the bubble surface near the bottom surface of the specimen. Direct observation confirmed that the flow pattern of the slag film along the rectangular specimen of the magnesia–chrome refractory in the local corrosion zone at the slag surface is similar to that shown in figure 3. Then, the interfacial tension difference is mainly caused by the concentration difference of MgO in the slag due to the dissolution of MgO from the refractory specimen into the slag.

(*b*) *The blast furnace trough material–(CaO–Al$_2$O$_3$–SiO$_2$) slag system (Mukai et al. 1984c, 1993)*

SiC granules on the surface of the trough material are oxidized by oxygen supplied from atmospheric gas via the slag film formed on the specimen of the material above the slag level, and are changed into carbon granules and SiO$_2$.

The dissolved SiO$_2$ from the trough material decreases the surface tension of CaO–Al$_2$O$_3$–SiO$_2$ slag (Mukai & Ishikawa 1981), which induces the slag film motion due to the Marangoni effect and the change in σ/ρ of the slag film. Slag film motion of this system is shown in figure 7. The flow mode of the meniscus was clarified by direct observation. Suspension of the carbon granules into the film and the slag film motion will facilitate the dissolution and abrasion of the trough material. Then, the local corrosion of this system arises only in an oxidizing atmosphere in the narrow zone of the trough material just above the slag level, though the slag film is also formed in an Ar atmosphere. Depth of the local corrosion increases in a step-like manner with time. This mode is analogous to that of the MgO(s)–(CaO–Al$_2$O$_3$–SiO$_2$) slag system (Harada *et al.* 1984). The limiting step of the corrosion rate for the rate retardation period is also considered to be the transport process of the accumulated SiO$_2$ from the slag surface into the bulk slag.

(*c*) *The blast furnace trough material–Fe–C alloy(l)–(CaO-Al$_2$O$_3$–SiO$_2$) slag system (Mukai et al. 1984d; Yoshitomi et al. 1986, 1987)*

The Marangoni flow of the slag film in the local corrosion zone of a metal–slag–trough material system is also induced by the concentration gradient of the slag film at the interface between the film and the metal in the vertical direction. The concentration gradient is caused by several kinds of reaction between the slag film and metal, or the film and the trough material, as shown in figure 10. The reaction (3.1) partially participates in the promotion of local corrosion by agitating the slag film as CO bubbles evolve:

$$(\text{SiO}_2) + 2\underline{\text{C}} \rightarrow \underline{\text{Si}} + \text{CO(g)}. \tag{3.1}$$

Carbon particles generated by reaction (3.2) are suspended in the slag film and dissolve into the metal phase when the carbon concentration in the metal is low, thus accelerating local corrosion:

$$\text{SiC(s)} + 2(\text{FeO}) \rightarrow (\text{SiO}_2) + 2\text{Fe(l)} + \text{C(s)}. \tag{3.2}$$

For metals containing high carbon content (e.g., in the vicinity of carbon saturation), suspended carbon particles remain in the slag film, which prevents motion of the film and reduces the contact area between the film and the metal, resulting in the reduction of the local corrosion rate.

210 K. Mukai

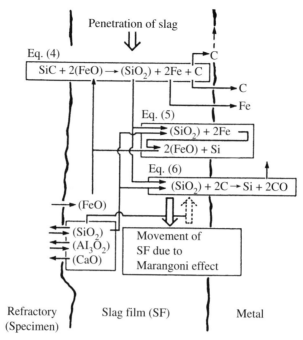

Figure 10. Transfer paths of reactant and products during the progress of local corrosion of trough materials at the slag–metal interface (Yoshitomi *et al.* 1987).

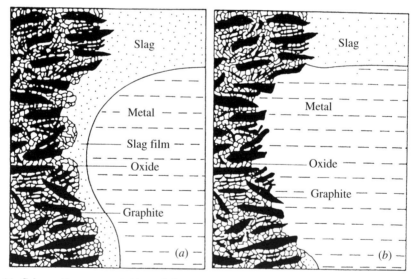

Figure 11. Schematic representation of the manner in which local corrosion of the immersion nozzle at the slag–metal interface proceeds (Mukai *et al.* 1989b).

(d) Al$_2$O$_3$–C *(Mukai et al. 1986b, 1989b)* or MgO–C *(Kii et al. 1995)*
refractories–Fe(l)–(CaO-Al$_2$O$_3$–SiO$_2$) *slag system*

As shown in figure 11, when the wall of the Al$_2$O$_3$–C or MgO–C refractory is initially covered with a slag film (figure 11a), the film not only wets the oxides, but dissolves them in preference to graphite. This changes the interface to a graphite-rich layer. Since the metal phase wets graphite better than the slag, the metal phase creeps up the surface of the specimen, as indicated in figure 11*b*, and dissolves graphite in preference to the oxides. Once the graphite-rich layer disappears due to dissolution into metal, the slag can again penetrate the boundary between the metal and the specimen, and the process is repeated. This cycle produces a local corrosion zone at the metal–slag interface. The up-and-down motion of the slag–metal interface shown in figure 11 was clearly observed using X-ray radiographic techniques.

The Marangoni flow of the slag film is also considered to play an important role in the local corrosion of immersion nozzles at the metal–slag interface during the stage shown in figure 11*a*.

4. Concluding remarks

Local corrosion of refractories at the slag surface is essentially caused by the active motion of the slag film formed by the wettability between the refractory and slag, since the slag film motion accelerates the dissolution rate of the refractory and also induces the abrasion of some refractories. The active film motion is dominantly induced by the Marangoni effect and/or change in the form of the slag film (slag meniscus) due to the variation of the surface tension and the density of the slag film.

The local corrosion of refractories at the slag–metal interface is also explained reasonably by a mechanism which is similar to that of the refractory–slag system. Then, the action of the surface (or interfacial) tension which leads to the local corrosion is very bad.

It is understood from this paper that the slag film of the $SiO_2(s)$–$(PbO$–$SiO_2)$ slag system moves with a mechanism similar to the 'tears of strong wine' mechanism clarified by Thomson (1855) more than 130 years ago. The trivial phenomenon of 'wine's tear' holds the key to the solution of the industrially serious technological problem of the local corrosion of refractories.

References

Brückner, R. 1967 *Glastech. Ber.* **40**, 451.

Caley, W. F., Marple, B. R. & Masson, C. R. 1981 *Can. Metall. Q.* **20**, 215.

Danckwerts, P. V. 1951 *Ind. Eng. Chem.* **43**, 1460.

Harada, T., Fujimoto, S., Iwata, A. & Mukai, K. 1984 *J. Jap. Inst. Metals* **48**, 181.

Hino, M., Ejima, T. & Kameda, M. 1967 *J. Jap. Inst. Metals* **31**, 113.

Iguchi, Y., Yurek, G. J. & Elliott, J. F. 1979 *Proc. 3rd Int. Iron and Steel Cong. ASM*, p. 346.

Jebsen-Marwedel, H. 1956 *Glastech. Ber.* **29**, 233.

Kii, T., Hiragushi, K., Yasui, H. & Mukai, K. 1995 *Unified Int. Technical Conf. on Refractory, 4th Biennial Worldwide Conf. on Refractories, Japan*, p. 379.

King, T. B. 1964 *Trans. Met. Soc. AIME* **230**, 1671.

Mukai, K. & Ishikawa, T. 1981 *J. Jap. Inst. Metals* **45**, 147.

Mukai, K., Iwata, A., Harada, T., Yoshitomi, J. & Fujimoto, S. 1983 *J. Jap. Inst. Metals* **47**, 397.

Mukai, K., Nakano, T., Harada, T., Yoshitomi, J. & Fujimoto, S. 1984*a Proc. 2nd Int. Symp. on Metallurgical Slags and Fluxes*, p. 207. Pennsylvania, Met. Soc. AIME.

Mukai, K., Masuda, T., Gouda, K., Harada, T., Yoshitomi, J. & Fujimoto, S. 1984*b J. Jap. Inst. Metals* **48**, 726.

Mukai, K., Masuda, T., Yoshitomi, J., Harada, T. & Fujimoto, S. 1984*c Tetsu-to-Hagane* **70**, 823.

Mukai, K., Yoshitomi, J., Harada, T., Hurumi, K. & Fujimoto, S. 1984*d Tetsu-to-Hagane* **70**, 541.

Mukai, K., Harada, T., Nakano, T. & Hiragushi, K. 1985 *J. Jap. Inst. Metals* **49**, 1073.

Mukai, K., Harada, T., Nakano, T. & Hiragushi, K. 1986*a J. Jap. Inst. Metals* **50**, 63.

Mukai, K., Toguri, J. M. & Yoshitomi, J. 1986*b Can. Metall. Q.* **25**, 265.

Mukai, K., Gouda, K., Yoshitomi, J. & Hiragushi, K. 1989*a Proc. 3rd Int. Conf. on Molten Slags and Fluxes*, p. 215. London: The Institute of Metals.

Mukai, K., Toguri, J. M., Stubina, N. M. & Yoshitomi, J. 1989*b ISIJ Int.* **29**, 469.

Mukai, K., Ikeda, E. & Yu, Z. 1993 *1st Int. Conf. on Processing Materials for Properties, Hawaii*, p. 273.

Schulte, K. 1977 *Glastech. Ber.* **50**, 181.

Sendt, A. 1965 *7th Cong. Int. du Verre, Bruxelles*, 352.1.

Thomson, J. 1855 *Phil. Mag.* **10**, 330.

Vago, E. & Smith, C. E. 1965 *Proc. 7th Int. Congress Glass, Brussels*, vol. II, .1.2, 62.1-22.

Yoshitomi, J., Harada, T., Hiragushi, K. & Mukai, K. 1986 *Tetsu-to-Hagane* **72**, 411.

Yoshitomi, J., Hiragushi, K. & Mukai, K. 1987 *Tetsu-to-Hagane* **73**, 1535.

Yu, Z. 1993 Ph.D. thesis, Kyushu Institute of Technology, pp. 15–74, 145.

Yu, Z. & Mukai, K. 1992 *J. Jap. Inst. Metals* **56**, 1137.

Yu, Z. & Mukai, K. 1995 *J. Jap. Inst. Metals* **59**, 806.

Zainan, T., Mukai, K. & Ogata, M. 1996 *Taikabutsu* **48**, 568.

Modelling of Marangoni effects in electron beam melting

By P. D. Lee[1], P. N. Quested[2] and M. McLean[1]

[1] Department of Materials, Imperial College of Science,
Technology and Medicine, Prince Consort Road, London SW7 2BP, UK
[2] Centre for Materials Measurement and Technology,
National Physical Laboratory, Teddington TW11 0LW, UK

Electron beam melting processes exhibit large thermal gradients in the region where the electron beam intercepts the melt; this leads to variations in the surface energy of the melt close to the beam inducing thermocapillary (Marangoni) flow. During melt processing of many materials the Marangoni contribution can dominate the fluid flow, influencing the trajectories of inclusions within the melt and providing a potential mechanism for controlling the removal and/or distribution of inclusions. A model of the macroscopic fluid flow and heat transfer, incorporating Marangoni effects, during electron beam melting has been developed and validated against surface flow observations during the electron beam button melting (EBBM) of IN718. The model indicates, and experimental observation confirms, that fluid flow in the molten pool is dominated by thermocapillary (Marangoni) forces, for the scale and operating conditions of the EBBM process. It is, therefore, possible to reverse the fluid flow through modification of the surface energy.

The effect of altering the concentration of sulphur, which is a highly surface active element, upon the Marangoni flow was determined both experimentally and computationally. The implications of altering this concentration on the effectiveness of inclusion removal and final material quality are discussed.

Keywords: Marangoni flow; superalloys; solidification modelling;
electron beam button melting

1. Introduction

The use of secondary melting processing, including electron beam techniques, is attracting increasing interest as the demands for cleaner materials for advanced engineering applications and recycling of materials grow. Examples include the use of electron beam cold-hearth refining for the production of titanium alloys for aero-engine parts which are highly stressed (Tilly et al. 1997); recycling of expensive nickel-based superalloys for turbine blades (Lowe 1994); purification of titanium metal (Tilmont & Harker 1996); the production of stainless steel for gas lines in the semiconductor industry (Lowe 1994) and electron beam drip melting for the production and purification of refractory metals (Schiller et al. 1982). The electron beam button melting (EBBM) test has been developed to evaluate the quality or cleanness of nickel-based alloys for turbine disc and blade applications (Shamblen et al. 1983; Sutton 1986; Quested & Hayes 1994). It provides a convenient well characterized small-scale electron beam melting procedure on which to validate our generic secondary melting modelling and to demonstrate the significance of Marangoni flow.

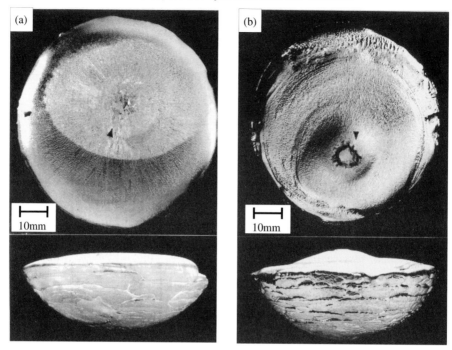

Figure 1. Comparison of EB melted button shape and cap of (*a*) VIM/VAR stock to (*b*) VIM/ESR stock for melts of IN718 under the same process conditions.

Electron beams provide a localized high-energy source that are able to heat a surface very rapidly to high temperatures. This has a number of consequences including: the production of high local temperature gradients; the evaporation of elements with high vapour pressures; and fusing of inclusions at the melt surface. To avoid these detrimental effects, high scan frequencies (approximately 1 kHz) of the beam on the material surface are used to minimize the dwell time. The effects of frequency have been summarized by Powell *et al.* (1995).

(1) At very high frequencies, dwell times are short and temperature fluctuations small.

(2) At high frequencies the temperature fluctuations are enough to affect the evaporation rates though they do not affect transient fluid flow.

(3) At moderate frequencies Marangoni flow is sufficient to modify temperature fluctuations and affect evaporation rates.

(4) At low frequencies other phenomena affect surface temperature, such as turbulent flow and depressions in the surface generated by large vapour pressure changes.

The EBBM test was developed because conventional metallographic evaluation techniques are inappropriate for the characterization of sparse inclusion distributions in superclean materials. It should be noted that as the number of inclusions in these materials decreases, larger volumes of alloy must be examined to obtain results which have statistical significance (Chone 1978). The electron beam button melting test has received much attention as a method for determining both the number, size and compositions of the inclusions which are concentrated from about a kilogram of alloy into a small 'raft' (*ca.* 5 mm diameter × 10 μm depth) on the surface of the solidified button. This technique has the advantage that it can handle volumes which are large enough to produce statistically significant results, and EBBM has proved useful in ranking the cleanness of superalloys. However, it has been found that in certain casts there is no apparent raft formation. In figure 1, the buttons obtained from two casts of IN718 prepared by vacuum induction melting/vacuum arc remelting (VIM/VAR) (see figure 1a) and VIM/electro-slag remelting (VIM/ESR) (see figure 1b) routes are compared and it can be seen that the VIM/VAR button has a flat surface and a poorly defined cap, whereas the VIM/ESR button has a well-defined cap and 'humped' top surface. In order to evaluate cleanness by EBBM, it is a prerequisite that a cap should be formed; consequently, in order to handle materials such as the VIM/VAR cast it is necessary to have a full understanding of the factors affecting raft formation. Moreover, if the reasons for these differences during EBBM processing can be understood, they will provide an insight into the larger scale electron beam manufacturing processes.

Dominique *et al.* (1984) have suggested the shape of the EBBM button, including raft formation, was affected by several factors including cleanness level, tramp element concentration (especially Cu) and the prior-melting processing. Analogous 'humping' of the top surface has been reported in TIG/GTA welding and it has been suggested that it is associated with the direction of the fluid flow in the molten metal pool (Mills & Keene 1990). There are close parallels between the EBBM and TIG/GTA welding processes that make it appropriate to investigate the role of surface flow direction on raft formation. In the liquid metal pool during localized electron beam heating, there are four principal forces affecting the fluid flow: buoyancy; electromagnetic (Lorentz); aerodynamic drag; and thermocapillary (Marangoni) forces. The last is usually dominant in the weld pool (Heiple & Roper 1982). Thermocapillary forces result from a surface tension gradient (in response to a temperature gradient) along the surface of the molten pool and surface flow occurs from regions of low to high surface tension. These surface flows then set off circulation flows within the molten pool.

The direction and magnitude of the thermocapillary flow in the molten pool are determined by the concentration of surface-active elements (especially the group VI elements O, S, Se and Te) present in the alloy. These elements cause a sharp temperature-dependent reduction in surface tension (Sahoo *et al.* 1988). The addition of these elements can significantly alter the dependence of surface energy on temperature ($\partial\gamma/\partial T$), often altering its sign from negative (for pure materials) to positive in the presence of highly surface active solutes. It is this latter effect which is responsible for the magnitude and direction of the fluid flow which are related to the Marangoni number (Ma) defined as

$$Ma = \frac{\partial\gamma}{\partial T}\frac{\partial T}{\partial x}L^2 \bigg/ \mu\alpha, \qquad (1.1)$$

where ($\partial T/\partial x$) is the temperature gradient along the surface, μ and α are the dynamic viscosity and thermal diffusivity of the liquid metal, and L is a characteristic length (e.g. the radius of the pool).

In pure metals and alloys with low concentrations of soluble O, S, Se and Te, $\partial\gamma/\partial T$ is negative and thus the surface tension will be highest in the cooler regions at the periphery of the pool (figure 2a) and thermocapillary flow will occur in a radially outward direction (Mills & Keene 1990; Heiple & Roper 1982). However, when the S or O concentration exceeds a certain 'critical' value (*ca.* 10 ppm at 1600 °C in Fe and probably in Ni-based alloys) (Mills & Keene 1990; Sahoo *et al.* 1988) the coefficient ($\partial\gamma/\partial T$) becomes positive. It can be seen from figure 2b that the thermocapillary

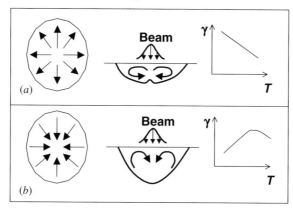

Figure 2. Typical pool flows: (a) low concentration of group VI elements; (b) concentration of group VI elements above critical level.

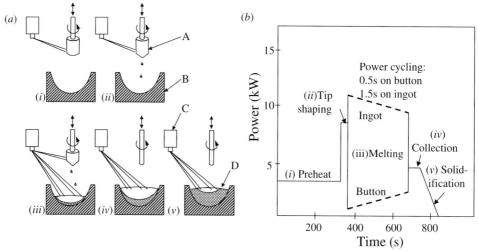

Figure 3. (a) Schematic diagram of the five stages of EBBM melting and (b) the associated power levels at each stage. The EBBM components are: A, ingot; B, copper crucible; C, electron beam source; D, button.

forces produce a radially inward flow. Note that the temperature gradients in molten pools are smaller in EBBM (ca. 10 K mm^{-1}) than in GTA welding (ca. 100 K mm^{-1}).

The objective of this research investigation was thus to determine whether raft formation is related to the direction of fluid flow in the molten pool and if so, to identify which force has a dominant effect on the fluid flow. The experimental methods will first be presented and the results of experiments to determine how changes in the concentrations of surface-active alloying additions alter the surface flow and final button structure will be described. A mathematical model of fluid flow, heat transfer and solidification, incorporating the effect of Marangoni forces, will be formulated. The results of the macroscopic process model, coupled to micromodels for structure prediction, will be presented and related to the fluid flow and the solidification structures observed experimentally on buttons in which the Marangoni forces have been modified by compositional control.

2. Experimental methods

The EBBM unit (Leybold AG ESI/07/30B) had a maximum emissive power and current of 30 kW and 1.2 A, respectively, and an accelerating voltage of 25 kV(6). A computer was incorporated into the instrument to control (i) the power (i.e. accelerating voltage and beam current), (ii) the position of the beam, (iii) the focusing of the beam and (iv) the positioning of the electrode in the furnace.

The button is formed by electron beam melting of a cylindrical ingot into a copper cooled hemispherical crucible of 168 mm diameter. The stages of the procedure are summarized by the steps shown schematically in figure 3a, with the accompanying typical power levels in figure 3b.

Two types of run were investigated. The first type was designed to allow characterization of the surface flow velocities. The second type was the standard run designed to give a controlled solidification pattern that concentrates inclusions into a central raft for characterization. The two types of run differ significantly only in the final stage. The key factors for the two runs are as follows.

(1) Stationary beam (flow characterization).

(i) A central beam was used to remelt a button after Al_2O_3 particles were added as tracers.

(ii) Surface flow velocities determined by measuring the particle motion during the EBBM process.

(2) Raft-forming solidification sequence (standard run).

(i) Solidification cycle of the beam going from the edge to the centre in an inwards radial motion whilst circumscribing high-frequency circumferential sweeps.

(ii) Inclusions are concentrated in a final central raft.

The chemical compositions of the two heats of IN718 used in the first experiments are given in table 1 as low-S and high-S. The alumina and NiS used in the doping trials described later were sieved to 43–63 µm; the reported purity of the NiS was 99.9%.

The buttons from the solidification sequence runs were examined metallographically after casting, to characterise the microstructural features. The secondary dendrite arm spacing measurements, were obtained from longitudinal sections of the button which were polished and then etched in Marble's reagent. Using an SEM, the length of five consecutive secondary arms was measured and averaged to determine the spacing. The primary dendrite arm spacing was measured using a mean linear intercept method from back-scattered electron micrographs.

3. Experimental results

(a) Stationary beam runs

The stationary beam runs were performed using identical conditions but with two different compositions, low-S and high-S (as given in table 1). For the low-S alloy, the surface flows were observed to be outwards from the centre of the beam with the particles reaching a maximum velocity of approximately 0.06 m s^{-1}. The particles travelled near to the edge of the pool, but were not transported all the way to the solid, indicating a small inward surface flow in the outermost region of the pool. The form of the flow is shown both schematically and by a still video frame in figure 4a. The Al_2O_3 particles are visible at the edge of the pool in the still video frame after rapidly flowing there from the centre.

Table 1. *Nominal composition of the two heats of IN718*

element	low-S wt%	high-S wt%	element	ppm	ppm
Ni	54.0	52.5	Mn	900	600
Fe	17.3	balance	S	6	20
Co	0.27	0.56	O	< 10	8
Cr	18.4	17.6	Se	< 3	—
Nb	5.1	5.0	Te	< 0.5	—
Mo	2.9	2.9	Ca	< 10	—
Ti	1.01	1.09	Mg	56	< 5
Al	0.52	0.52	N	90	87
C	0.033	0.038			

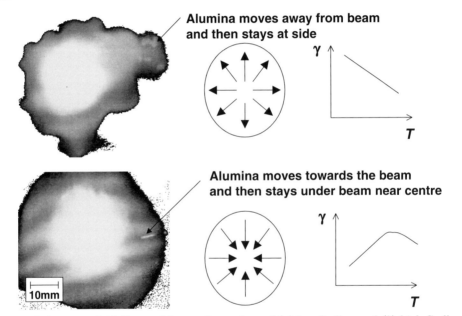

Figure 4. Behaviour of Al_2O_3 particles on the surface of (*a*) low-S alloy and (*b*) high-S alloy as shown by a video still. The schematic diagrams at the right-hand side show the flow patterns that would produce such behaviour.

For the high-S experiments, the particles were observed to move inwards from the edge of the pool with peak velocities of 0.19 m s^{-1}, moving at highest velocity shortly after leaving the edge of the pool and slowing to velocities of approximately 0.1 m s^{-1} half way towards the centre. The particles stop before reaching the centre, on the inside of the circle that the electron beam describes. The video frame in figure 4*b* clearly shows an Al_2O_3 particle moving quickly from the edge of the pool into the centre, confirming the flow patterns shown schematically in the same figure.

The results therefore indicate that the change in minor alloying elements between the low-S and high-S compositions is sufficient to reverse the direction of fluid flow. If raft formation is primarily controlled by the direction of the thermocapillary flow it would be expected that for melts that currently show no clear raft formation (e.g. VIM/VAR) a raft could be induced by modifying the sulphur concentration to produce a radially inward flow. To test this hypothesis, an additional run was made using the low-S composition but NiS powder was added in the same manner as the Al_2O_3 particles, increasing the sulphur level to 100 ppm. The flow observed was inwards, similar to that observed for alloy high-S.

(b) Raft-forming solidification sequence runs

The microstructural features (primary and secondary dendrite arm spacing) of electron beam melted buttons were measured on sectioned buttons undergoing a *controlled solidification* stage during processing. During this stage the electron beam moves inwards whilst the power is reduced. The results are shown in figures 10a and 11a where they are compared to the model results as discussed later.

4. Mathematical model formulation

(a) Macro-model

Electron beam melting is a complex process involving many different phenomena, all of which have to be incorporated in the model in order to produce a realistic prediction of the interaction of the process parameters upon the final ingot structure. The model must therefore incorporate: heat transfer; fluid flow; magnetohydrodynamics, including both the Lorentz force and Joule heating; Marangoni forces; and solidification. The macro-modelling was performed using the commercial finite volume code *Fluent*[TM] (registered by Fluent, Inc.) with subroutines added to handle the spatially and temporally varying heat flux (both from the electron beam and the heat transfer to the crucible) and surface tension gradient boundary conditions. Magneto-hydrodynamic effects were also coded in user subroutines but were found to be negligible for the cases studied, and hence are not included here. The equations solved are listed below.

(b) Energy equation

The energy equation, assuming incompressible flow, is

$$\frac{\partial}{\partial t}(\rho h) + \nabla \cdot (\rho \boldsymbol{u} h - k \nabla T) = S_h, \tag{4.1}$$

where h is the static enthalpy, ρ is the density, k is the thermal conductivity, t is time, $\boldsymbol{u}$ is the fluid velocity and T is temperature. S_h represents heat sources, i.e. electron beam heating (EBBM), and energy gains/losses by radiation, conduction and latent heat evolution.

(c) Momentum and continuity equations

The velocity of liquid, $\boldsymbol{u}$, is given by the momentum equation (Hirsch 1988):

$$\frac{\partial}{\partial t}(\rho \boldsymbol{u}) + \nabla \cdot (\rho \boldsymbol{u} \otimes \boldsymbol{u} + p\boldsymbol{I} - \tau) = \rho \boldsymbol{g} + \boldsymbol{F}, \tag{4.2}$$

where $\boldsymbol{I}$ is the 3×3 identity tensor, $\boldsymbol{g}$ is the gravitational acceleration, and $\boldsymbol{F}$ is the sum of all other body forces (e.g. the Marangoni force or the Lorentz force).

The stress tensor τ has components

$$\tau_{ij} = \left[\mu \left(\frac{\partial u_i}{\partial x_j} + \frac{\partial u_j}{\partial x_i} \right) \right] - \frac{2}{3} \mu \frac{\partial u_l}{\partial x_l} \delta_{ij}, \tag{4.3}$$

where μ is the viscosity.

To solve for the momentum and pressure, the mass conservation equation must also be satisfied:

$$\frac{\partial \rho}{\partial t} + \nabla \cdot (\rho \boldsymbol{u}) = S_{\mathrm{m}}. \tag{4.4}$$

S_{m} is the mass added to the continuous phase from any dispersed phase or as a source (e.g. due to droplets from the consumable electrode). In the present work, S_{m} has been taken to be zero, since filling is not modelled.

(d) Micro-models

To illustrate how the Marangoni induced flow patterns affect the final microstructure, several micro-models were written to predict aspects of the final solidification structure.

As summarized by McLean (1983) the primary dendrite arm spacing, λ_1, in superalloys has been shown both experimentally and theoretically by many authors to be a function of the thermal gradient and the growth velocity. Recently Lu & Hunt (1992) developed a numerical model of cellular and dendritic growth to predict the cell and dendrite spacing as well as the undercooling at the tip. Hunt & Lu (1995) provided an analytical expression fitted to their numerical model results that predicts the minimum stable half-spacing of dendrites. Cast in terms of non-dimensional parameters†, Hunt & Lu found for the case of $G' > 1 \times 10^{-10}$ and $0.068 < k < 0.69$ that the dimensionless dendrite half spacing, λ', is given by

$$\lambda' = 0.7798 \times 10^{-1} V'^{(a-0.75)} (V' - G')^{0.75} G'^{-0.6028}, \tag{4.5}$$

where

$$a = -1.131 - 0.1555 \log(G') - 0.7589 \times 10^{-2} [\log(G')]^2. \tag{4.6}$$

To determine λ_1 the thermal gradient and front velocity were calculated using the macro-model. As a cell passes through the liquidus temperature the nearest neighbours' temperatures were used to calculate the thermal gradient normal to the isotherm. Hunt & Lu specify G as the solid thermal gradient, referring to the dendrite tip as the solid. In the macro model the mushy zone is treated as a continuum, making the thermal gradient close to the liquidus temperature an appropriate approximation for G. In addition for equation (4.5) to be applied, the superalloy being studied was approximated as a pseudo-binary alloy.

The value of V used to determine λ_1 was the velocity of the liquidus isotherm, V_{liq}. This velocity was determined using a central difference approximation of the following derivative:

$$V_{\mathrm{liq}} = \left. \frac{\partial n}{\partial t} \right|_{T=T_{\mathrm{liq}}} = 1 \left/ \left. \frac{\partial t}{\partial n} \right|_{T=T_{\mathrm{liq}}} \right. = 1 \left/ \frac{\partial t_1}{\partial n} \right. , \tag{4.7}$$

where t_1 is the time at which the local temperature attains the liquidus value and n is the direction normal to the liquidus isotherm.

---

† The non-dimensional parameters are

$$G' = \frac{G\Gamma k}{\Delta T_0^2}, \quad V' = \frac{V\Gamma k}{D\Delta T_0}, \quad \lambda' = \frac{\lambda \Delta T_0}{\Gamma k}, \quad \Delta T_0 = \frac{mC_0(k-1)}{k},$$

where ΔT_0 is the undercooling for a planar front and G, V, Γ, D, m, C_0 and k are, respectively, the solid temperature gradient, velocity, Gibbs–Thomson coefficient, liquid diffusion coefficient, liquidus slope, bulk composition and distribution coefficient (Hunt & Lu 1995).

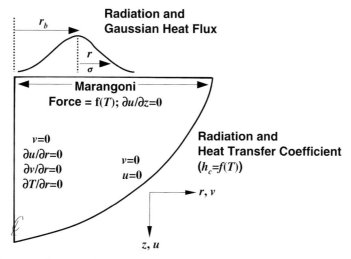

Figure 5. Schematic diagram showing the geometry and boundary conditions used to model the EBBM process.

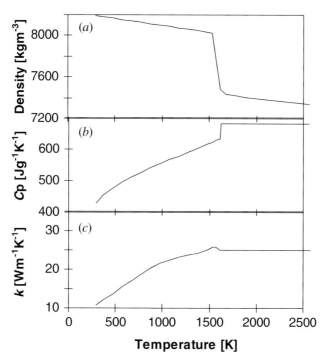

Figure 6. The temperature-dependent material properties used in the EBBM simulations: (*a*) density; (*b*) specific heat capacity; (*c*) thermal conductivity.

The ripening or local solidification time t_s, is defined as the time for which the dendrite is in the mushy zone. The secondary dendrite arm spacing, λ_2, is related to t_s by

$$\lambda_2 = -7.0 + 12.5 t_s^{0.33} \quad (\mu m). \tag{4.8}$$

Table 2. *Values used to simulate EBBM processing of IN718*

(Note that $f(t)$ indicates the value is a function of time whilst $f(T)$ indicates a function of temperature.)

property	symbol	value	units
button radius	r	37	mm
button depth	z	25	mm
beam current	I	$f(t)$	mA
beam voltage	V	25	kV
beam focal radius	r_σ	10	mm
beam location radius	r_b	$f(t)$	mm
density	ρ	$f(T)$	$\mathrm{kg\ m^{-3}}$
specific heat capacity	C_p	$f(T)$	$\mathrm{J\ kg^{-1}\ K^{-1}}$
viscosity	ν	5×10^{-3}	$\mathrm{kg\ m^{-1}\ s^{-1}}$
surface tension gradient	$\partial\gamma/\partial T$	$f(T)$	$\mathrm{N\ m^{-1}\ K^{-1}}$
liquidus temperature	T_l	1609	K
solidus temperature	T_s	1533	K
latent heat	L	270 000	$\mathrm{J\ kg^{-1}}$
ingot/crucible	h_c	$f(T_{\mathrm{ingot}})$ $(T_{\mathrm{amb}} = 500\ \mathrm{K})$	$\mathrm{W\ m^{-2}}$
ingot/crucible emissivity	ε	0.3 $(T_{\mathrm{amb}} = 500\ \mathrm{K})$	—
button top emissivity	$\varepsilon_{\mathrm{top}}$	0.25 $(T_{\mathrm{amb}} = 273\ \mathrm{K})$	—

(e) Problem formulation

The heat transfer and fluid flow was modelled in the EBBM assuming that the flow and distribution of heat from the electron beam were axisymmetric. The model was solved on a grid of 40×40 control volumes for transient flow using an implicit solution with time steps of 0.5 s. A steady-state solution was used as an initial condition assuming a highly defocused beam centred half way out the radius. The geometry and boundary conditions used are shown in figure 5. The material properties used are listed in table 2 or plotted as a function of temperature in figure 6.

The values for $\partial\gamma/\partial T$ were estimated using values provided by Mills (1995, personal communication) and assuming the behaviour of the nickel-base alloy IN718 is similar to that of sulphur in Fe–Ni–Cr alloys, as determined by McNallan & Debroy (1991). The values were included as a piece-wise linear fit. The values used are shown in figure 7a for the two cases of low surfactant concentration (6 ppm S, < 10 ppm O), and high surfactant concentration (20 ppm S, 8 ppm O), the low-S and high-S compositions of IN718 as given in table 1, respectively.

The values for the heat transfer coefficient between the ingot and mould wall, h_c, were calculated from measurements of the heat flux into a copper crucible made during the plasma remelting of IN718 into a 125 mm diameter cylindrical ingot as given by Lothian *et al.* (1997). The heat flux was divided into radiative, convective, and contact components, with the h_c value representing the convective and contact portion, whilst a value of $\varepsilon = 0.3$ was used to calculate the radiative component.

Starting with the steady state solution, the process was modelled with the tran-

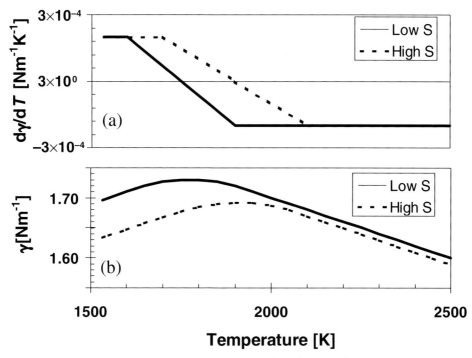

Figure 7. The temperature-dependent values for (a) $\partial\gamma/\partial T$ and (b) γ used in the EBBM simulations.

sient boundary condition of the electron beam moving across the surface providing a Gaussian distribution of heat flux, Q, characterized by

$$Q(R) = Q_0 e^{-R^2/r_\sigma^2}, \qquad (4.9)$$

where R is the distance from the beam centre, r_σ is the beam focal radius, and Q_0 is the total flux. Given that the beam circumscribes the centre of the button at a radius of r_b, the circumferentially averaged heat flux (i.e. averaged over one full sweep of θ), $Q_\sigma(r)$, can be obtained by integrating Q by $\mathrm{d}\theta$, giving

$$Q_\theta(r) = \frac{1}{\pi} \int_0^\pi Q(R)\,\mathrm{d}\theta$$

$$= Q_0 e^{-(r^2+r_b^2)/r_\sigma^2} I_0\left(\frac{2rr_b}{r_\sigma^2}\right), \qquad (4.10)$$

where I_0 is the modified Bessel function of the first kind and order zero.

5. Model results and discussion

(a) Stationary beam simulations

Using the model outlined in the previous section, the two stationary beam experimental runs were simulated, both with the same thermal boundary conditions but with the low and high sulphur content being represented by the two expressions for $\partial\gamma/\partial T$ as a function of T shown in figure 7a. The electron beam motion was the same for both cases:

(i) 30 s of r_b varying from 25 to 5 mm over 2 s cycles at a power of 6 kW;

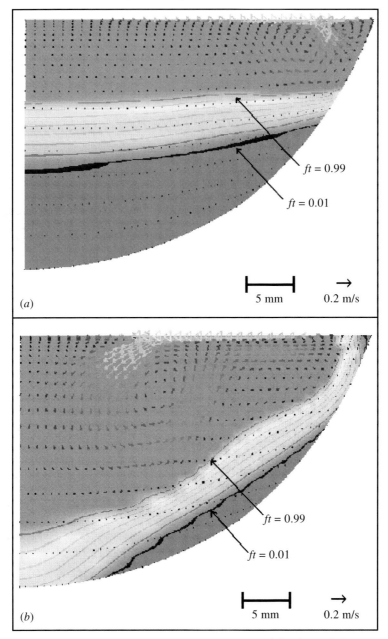

Figure 8. Predicted flow patterns using a 'central beam' with (a) a low sulphur content and (b) a high sulphur content.

(ii) 30 s with no heat flux (to simulate the time during which the Al_2O_3 particles were added);

(iii) and finally 60 s of r_b varying from 20 to 0 mm over 20 s cycles at a power of 6 kW.

For the low-S case the surface flow is predicted to be outwards from the centre of the beam with the particles reaching a maximum velocity of approximately $0.16\ \mathrm{m\,s^{-1}}$. The particles are assumed to be markers of the surface flow which is calculated to be from the centre to near to the edge of the pool; however, a small inward surface flow at the outermost region of the pool is predicted. The calculated flow pattern 78 s into the simulation (seconds after the pool surface had become fully molten) is shown in figure 8a. The predicted surface flow is outwards, which is consistent with the observation: however, the predicted flow velocity is significantly higher than the maximum velocities observed experimentally (0.06 m s^{-1}). In the experimental study it was not possible to track the Al_2O_3 particles in the high-velocity region. As observed experimentally, a small recirculating inward flow near the edge is predicted, with a time-dependent size and peak velocity. This flow is caused by the positive value of $\partial\gamma/\partial T$ at temperatures less than 1750 K, and the size of inward flow is a function of the location of this isotherm, which is in the time-dependent stage of remelting, and on the value of r_b, the size of the incident heat flux.

For the high-S simulation, the particles were observed experimentally to move inwards from the edge of the pool with peak velocities of 0.19 m s^{-1}, moving at highest velocity shortly after leaving the edge of the pool, slowing to velocities of approximately 0.1 m s^{-1} half way towards the centre. The particles stop before reaching the centre but on the inside of r_b. Figure 8b shows the predicted flow pattern 78 s into the simulation. This period is when the velocities were first recorded experimentally. The predicted surface flow pattern is identical to that observed experimentally, with the flow going from the outside into the centre, but slowing down just before reaching the centre. The maximum velocity predicted is 0.14 m s^{-1}, lower than that observed experimentally (0.19 m s^{-1}). This suggests that the value for $\partial\gamma/\partial T$ may be greater than that used in the simulation, or that the inversion point from a positive to a negative value could be at a higher temperature than the value of 1900 K.

Comparing the two cases, a reduction in the rate of change of $\partial\gamma/\partial T$ from a positive to negative value and a $150\,^{\circ}\text{C}$ increase in the inversion point, dramatically changed the flow patterns and location of the liquidus front. The size of the mushy zone is also altered, and hence the microstructural features will be different. A comparison of figures 8a, b illustrates the dominance of Marangoni flow in the EBBM process.

(b) Raft-forming solidification sequence runs

The microstructural features of electron beam melted buttons undergoing a *controlled solidification* stage during processing, as described in the experimental methods, were simulated. During the *controlled solidification* the electron beam moved inwards whilst the power was reduced. This process was modelled with r_b going from 35 mm to zero over 120 s whilst the power decays linearly from 2.6 kW to zero. The effectiveness of this raft-forming sequence for high sulphur levels is shown in figure 9 where the flow is shown at eight stages during stage (v) of the EBBM process. Note that the peak velocities drop quickly as the electron beam power is reduced giving a more quiescent flow into the centre that propels the inclusions into a central raft with less turbulence. The thermal conditions during this solidification sequence (and a simulation using the low-S properties) were used for predictions of the resulting microstructure.

The resulting predictions for λ_1 for low-S properties are shown in figure 10a calculated using equation (4.5) with the material properties for Γ, D, m, C_0 and k given in table 2. No predictions could be made in the bottom region of the button (cross-hatched area in figures 10 and 11) because this area was already mushy in

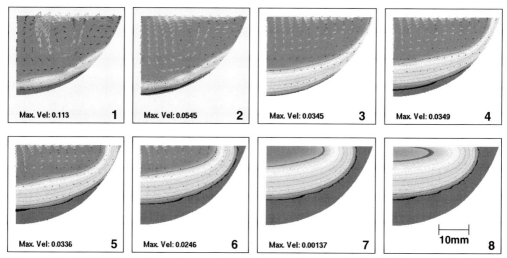

Figure 9. Predicted flow patterns at several times during the controlled solidification raft-forming stage (v) of the EBBM process for a high sulphur level (simulating high-S composition). Eight different times relative to the beginning of the controlled solidification stage are shown: 1, 0 s; 2, 10 s; 3, 26 s; 4, 51 s; 5, 76 s; 6, 101 s; 7, 116 s; 8, 121 s.

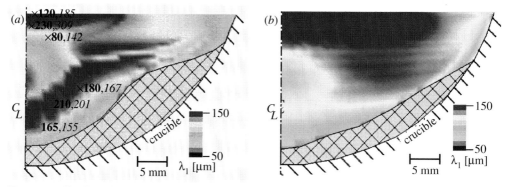

Figure 10. Predicted primary dendrite arm spacing for a button with a 120 s controlled solidification stage for (a) a low sulphur content and (b) a high sulphur content. For (a) experimentally measured values are superimposed in bold with the predicted values are in italics beside them.

the steady-state solution used as the initial condition for the model (i.e. the thermal history could not be tracked from the fully liquid state).

The values measured experimentally are listed in figure 10a (in bold) beside the predicted values (in italics). Near the top of the button at the centre line the model prediction fails. This is the region to solidify last and the only area where the gradients are so low that a liquidus isotherm is predicted to enter from the top of the button due to radiative heat loss competing with conduction through the button. (Note how quickly the isotherms progress in during the final stages of the solidification sequence shown in figure 9.) Experimentally this region is occasionally found to be equiaxed, indicating that the columnar dendrites can not grow in from the sides sufficiently quickly to prevent strong undercoolings. The λ_1 model assumes that the growth is near steady state, and this assumption does not appear to hold in this region. When the columnar dendrites were found to extend to the top, the spacing was smaller than predicted, suggesting that the dendrites could not adjust their

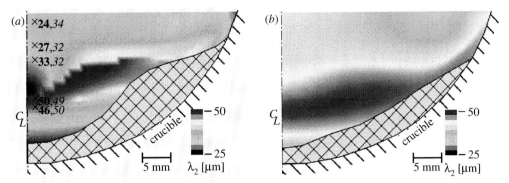

Figure 11. Predicted secondary dendrite arm spacing for a button with a 120 s controlled solidification stage for (*a*) a low sulphur content and (*b*) a high sulphur content. For (*a*) experimentally measured values are superimposed in bold with the predicted values are in italics beside them.

spacing in this relatively small distance. Pratt & Grugel (1993) have shown experimentally that the λ_1 adjust to order of magnitude changes, but not to relatively small changes, in withdrawal velocity during directional solidification experiments, suggesting either a slow response time to changes in thermal conditions or that the stable growth regime is large, adding a hysteresis effect.

The predictions for λ_2 for the low-S material are shown in figure 11*a* calculated using equation (4.8). The correlation of predicted values (italics) to experimental (bold) is good. The good agreement of both the λ_1 and λ_2 predictions to those measured indicates that the model is predicting the correct thermal histories and hence flow patterns.

The effect of altering only the value of $\partial\gamma/\partial T$, representing the change in sulphur composition, upon the final microstructure is illustrated by comparing the low-S and high-S predictions for λ_1 and λ_2 in figures 10*a* and 11*a* to figures 10*b* and 11*b*. The dominance of the Marangoni effect upon the fluid flow, and therefore the structure of the solidified IN718 buttons is clear. The Marangoni force alters the inclusion trapping (raft-forming) tendency of the process, the microstructure produced by the process and hence properties of the metal produced. These findings are specific to the operating conditions of the EBBM process used, and the conclusions can not be directly extended to large scale electron beam processing, such as electron beam cold-hearth refining. However, the model used in the present study can be extended to other manufacturing processes and the successful simulation of the well characterized EBBM process is a valuable validation of the computational approach used in the model.

6. Conclusions

The comparison of predicted and observed surface flows in electron beam melted buttons confirms that the Marangoni force is the main driving force for fluid flow in the EBBM process. Minor variations in the dependency of the surface tension on temperature can cause dramatically different flows, as shown by increasing the concentration of the surfactant sulphur from 6 to 20 ppm. Modelling of the EBBM process is in agreement with the experimentally based hypothesis, illustrating that a shift in the inversion temperature at which the value of $d\gamma/dT$ goes from a positive to negative value by only 150 °C can reverse the flow in the molten pool.

The modelling of the raft-forming solidification sequence runs was extended to predict aspects of the microstructures produced, and these predictions were validated by measurements of the primary and secondary dendrite arm spacings. The coupling of the macro-model of the fluid flow and heat transfer to microstructural models illustrated that the changes in flow, caused by the different driving forces, have a large impact on the final microstructural features of the superalloy. It is therefore critical to consider the Marangoni effect when processing metals using electron beam melting.

The authors thank a number of colleagues for their contributions to the work described in this paper: D. M. Hayes, K. C. Mills, R. M. Lothian and L. J. Hobbs. The authors also thank R. M. Ward and T. P. Johnson of the IRC in Materials for High Performance Applications, Birmingham, for providing the experimental data from which the heat transfer coefficients were determined. Part of the research reported in this paper was carried out as part of the 'Materials Measurement Programme', a programme of underpinning research at NPL financed by the United Kingdom Department of Trade and Industry. Aspects of the work were also supported by EPSRC Grant no. GR/J65068, the DERA, Farnborough, and INCO Alloys Ltd, Hereford, have provided material, data, and their insight.

References

Chone, J. 1978 Echantillonnage de billettes de coulee continue en vue de la description de la structure interne et la properte inclussionnaire. *Proc Symp. on Quantitative Metallography, Florence, Nov 1978*, publ. Ass. Italiana di Metallurgia, pp. 209–224.

Dominique, J. A., Sutton, W. H. & Yu, K. O. 1984 Characterization of VIM, VIM/VAR and VIM/ESR IN-718 EB-test buttons. *Proc. Conf. on Electron Beam Melting and Refining State of the Art 1984, Reno, NV* (ed. R. Bakish), pp. 330–346. Englewood Cliffs, NJ: Bakish Mat. Corp.

Heiple, C. R. & Roper, J. R. 1982 Mechanism for minor element effect on GTA fusion zone geometry. *Welding. J.* **61**, 97–102.

Hirsch, C. 1988 *Numerical computation of internal and external flows*, vol. 1. New York: Wiley.

Hunt, J. D. & Lu, S.-Z. 1995 Numerical modelling of cellular/dendritic array growth: spacing and undercooling predictions. *Modelling of casting, welding and advanced solidification processes VII* (ed. M. Cross & J. Campbell), pp. 525–532. Warrendale, PA: TMS.

Lothian, R., Lee, P. D., McLean, M., Ward, R. M., Johnson, T. P. & Jacobs, M. H. 1997 Modelling the liquid pool shape during plasma melting of turbine disc superalloys. *Proc. Int. Sym. on Liquid Metal Proc. & Casting, Santa Fe, New Mexico, 16/2–19/2/97* (ed. A. Mitchell & P. Auburtin), pp. 133–144. AVS.

Lowe, J. H. C. 1994 Electron-beam cold-hearth refining in Vallejo. *Proc. Conf. on Electron Beam Melting and Refining State of the Art 1994, Reno, NV* (ed. R. Bakish), pp. 69–77. Englewood Cliffs, NJ: Bakish Mat. Corp.

Lu, S.-Z. & Hunt, J. D. 1992 A numerical analysis of dendritic and cellular array growth: the spacing adjustment mechanisms. *J. Crystal Growth* **123**, 17–34.

McLean, M. 1983 *Directionally solidified materials for high temperature service*, pp. 28–33. London: The Materials Society.

McNallan, M. J. & Debroy, T. 1991 Effect of temperature and composition on surface tension in Fe–Ni–Cr alloys containing sulphur. *Met. Trans.* B **22**, 557–560.

Mills, K. C. & Keene, B. J. 1990 Factors affecting variable weld penetration. *Int. Mater. Rev.* **35**, 185–216.

Powell, A., Van Den Avyle, J., Damkroger, B. & Szekely, J. 1995 Simulation of multicomponenet losses in electron-beam melting and refining at varying scan frequencies. *Proc. Conf. on Electron Beam Melting and Refining State of the Art 1995, Reno, NV* (ed. R. Bakish), pp. 263–277. Englewood Cliffs, NJ: Bakish Mat. Corp.

Pratt, R. A. & Grugel, R. N. 1993 Microstructural response to controlled accelerations during the directional solidification of Al–6 wt.% Si alloys. *Mater. Charact.* **31**, 225–231.

Quested, P. N. & Hayes, D. M. 1994 The evaluation of cleanness by electron-beam button melting and other methods: a review. *Proc. Conf. on Electron Beam Melting and Refining State of the Art 1994, Reno, NV* (ed. R. Bakish), pp. 6–38. Englewood Cliffs, NJ: Bakish Mat. Corp.

Sahoo, P., Debroy, T. & McNallan, M. J. 1988 Surface-tension of binary metal: surface-active solute systems under conditions relevant to welding metallurgy. *Metall. Trans.* B **19**, 483–491.

Schiller, S., Heisig, U. & Panzer, S. 1982 *Electron-beam technology*, pp. 255–288. New York: Wiley.

Shamblen, C. E., Culp, S. L. & Lober, R. W. 1983 Superalloy cleanliness evaluation using the EB button melt test. *Proc. Conf. on Electron Beam Melting and Refining State of the Art 1983, Reno, NV* (ed. R. Bakish), pp. 61–94. Englewood Cliffs, NJ: Bakish Mat. Corp.

Sutton, W. H. 1986 Electron beam remelt, substance, scope and future as a quality control tool. *Proc. Conf. on Electron Beam Melting and Refining State of the Art 1986, Reno, NV* (ed. R. Bakish), pp. 297–317. Englewood Cliffs, NJ: Bakish Mat. Corp.

Tilly, D. J., Shamblen, C. E. & Buttrill, W. H. 1997 Premium quality Ti alloy production: HM + VAR status. *Proc. Int. Sym. on Liquid Metal Proc. & Casting, Santa Fe, NM, 16/2–19/2/1997* (ed. A. Mitchell & P. Auburtin), pp. 85–96. AVS.

Tilmont, S. & Harker, H. 1996 THT: an update. *Proc. Conf. on Electron Beam Melting and Refining State of the Art 1996, Reno, NV* (ed. R. Bakish), pp. 191–197. Englewood Cliffs, NJ: Bakish Mat. Corp.

Protein crystal movements and fluid flows during microgravity growth

By Titus J. Boggon[1], Naomi E. Chayen[2], Edward H. Snell[3],
Jun Dong[1,10], Peter Lautenschlager[4], Lothar Potthast[4],
D. Peter Siddons[5], Vivian Stojanoff[6], Elspeth Gordon[7],
Andrew W. Thompson[6,8], Peter F. Zagalsky[9], Ru-Chang Bi[10]
and John R. Helliwell[1]

[1]Structural Chemistry Section, Chemistry Department, University of Manchester,
Oxford Road, Manchester M13 9PL, UK
[2]Department of Biophysics, The Blackett Laboratory, Imperial College of Science,
Technology and Medicine, London SW7 2BZ, UK
[3]NASA, Laboratory for Structural Biology, Code ES76, Building 4464,
MSFC, Huntsville, AL 35812, USA
[4]Dornier GmbH, Raumfahrt-Infrastruktur, 88039 Friedrichshafen, Germany
[5]NSLS, Brookhaven National Laboratory, Upton, NY 11973, USA
[6]ESRF, BP 220, 38043 Grenoble Cedex, France
[7]Laboratoire de Cristallographie Macromoléculaire, Institut de Biologie Structurale,
41 Avenue des Martyrs, 38027 Grenoble Cedex, France
[8]EMBL, Ave. des Martyrs, 38043 Grenoble Cedex, France
[9]Biochemistry Department, Royal Holloway and Bedford New College,
University of London, Egham, Surrey TW20 0EX, UK
[10]Institute of Biophysics, Academia Sinica, Beijing 100101, China

The growth of protein crystals suitable for X-ray crystal structure analysis is an important topic. The methods of protein crystal growth are under increasing study whereby different methods are being compared via diagnostic monitoring including charge coupled device (CCD) video and interferometry. The quality (perfection) of protein crystals is now being evaluated by mosaicity analysis (rocking curves) and X-ray topographic images as well as the diffraction resolution limit and overall data quality. Choice of a liquid–liquid linear crystal-growth geometry and microgravity can yield a spatial stability of growing crystals and fluid, as seen in protein crystallization experiments on the uncrewed platform EURECA. A similar geometry used within the Advanced Protein Crystallization Facility (APCF) onboard the crewed shuttle missions SpaceHab-01 and IML-2, however, has shown by CCD video some lysozyme crystal movement through the mother liquor. Moreover, spurts and lulls of growth of a stationary lysozyme protein crystal that was probably fixed to the crystal-growth reactor wall suggests g-jitter stimulated movement of fluid on IML-2, thus transporting new protein to the growing crystal faces. In yet another study, use of a hanging drop vapour diffusion geometry on the IML-2 shuttle mission showed, a gain via CCD video monitoring, growing apocrustacyanin C_1 protein crystals exec iting near cyclic movement, reminiscent of Marangoni convection flow of fluid, the crystals serving as 'markers' of the fluid flow. These observations demonstrated that the use of vapour diffusion geometry did not yield spatially stable crystal position or fluid conditions for a solely protein diffusive regime to be realized. Indeed mosaicity evaluation of those vapour diffusion-grown apocrustacyanin C_1 crystals showed inconsistent protein crystal quality, although the best crystal studied was micro-gravity grown. In general, realizing perfect conditions for protein crystal growth, of absence of movement of crystal or fluid, requires not only the correct choice of geometry but also the avoidance of low-frequency ($\lesssim 5$ Hz) g-jitters. A review is given here of existing results and experience over several microgravity missions. Some comment is given on gel protein crystal growth in attempts to 'mimic' the benefits of microgravity on Earth. Finally, the recent new results from our experiments on the shuttle mission LMS are described. These results include CCD video as well as interferometry during the mission, followed, on return to Earth, by reciprocal space mapping at the NSLS, Brookhaven, and full X-ray data collection on LMS and Earth control lysozyme crystals. Diffraction data recorded from LMS and ground control apocrustacyanin C_1 crystals are also described.

Keywords: protein crystallization; microgravity; interferometry; CCD video;
crystal perfection; g-jitter; Marangoni convection

1. Background

The monitoring of crystal growth using CCD video allows key data to be collected on the process of crystallization, such as growth rates and movements of crystals (figure 1). Clearly then, crystallization geometries can be directly compared as well as the 'quality' of individual missions, along with Earth control conditions. Moreover, the state of the mother liquor can be monitored via interferometry, which measures refractive index changes which can be due to the flows of, for example, precipitating agents like salt, and subsequently gradual depletion of protein as crystals nucleate and grow.

Harvesting of crystals is made on return to Earth from a microgravity mission where a variety of X-ray analyses can be conducted, especially using high brilliance synchrotron X-ray sources in order to combine fine collimation and strong intensity to probe weak reflections. Hence, mosaicity measurements, as well as X-ray topographic images, can be collected, crystal quality compared and further insights gained into the various crystallization methods and their relative success.

2. Resumé of previous experiments onboard shuttle missions

Apocrustacyanin C_1 crystal growth in vapour diffusion geometry was conduct-ed on the IML-2 (STS-65), USML-2 (STS-73) and LMS (STS-78) shuttle missions. The experiments were monitored with CCD video observation (Chayen *et al.* 1997) and thereby provided some insights into the effects of Marangoni convection on the crystallization process in microgravity.

Lysozyme crystal growth has been conducted using the dialysis membrane tech-nique (in dialysis geometry crystallizations there is no gas–liquid phase boundary) provided by the APCF onboard the SpaceHab-01 (STS-57) and IML-2 (STS-65) as well as recently on the LMS (STS-78) shuttle missions. The APCF (Advanced Pro-tein Crystallisation Facility) is a multiuser facility built by Daimler–Benz Aerospace–Dornier under the contract of the European Space Agency ESA. Significant crystal movements were viewed on all these missions. However, in relation to the Marangoni convections described below these movements are generally slight, with one excep-tion. On SpaceHab-01, well into the mission, and after crystal growth had largely been completed, a sudden and dramatic motion of the crystals occurred attributed to the retrieval of the EURECA satellite (see below).

The CCD video in the APCF can view crystallizations using either a wide field-of-view (WFOV) or a narrow field-of-view (NFOV) lens (Bosch *et al.* 1992; Snyder *et al.* 1991). The lenses are non-interchangeable, and therefore a specified crystallization can be monitored only using either WFOV or NFOV optics. WFOV monitors the whole protein chamber (6.48×8.59 mm^2), whereas NFOV monitors only a magnified portion (3.77×4.99 mm^2). Due to the constraints (power limitation) placed upon the apparatus, recording of the digital images takes a period of about 4 min. This, combined with moving the camera to view other reactors, has restricted the frequency of images that can be obtained.

(a) *Apocrustacyanin C$_1$ on the IML-2 and USML-2 missions*

Apocrustacyanin C$_1$ has been crystallized on the IML-2, USML-2 and LMS shuttle missions. On IML-2, CCD video was used to follow the crystallization. Crystals grew in the vapour diffusion droplet and moved in a circular way, consistent with that of Marangoni convection (Chayen *et al.* 1997; Savino & Monti 1996) (figure 1*a*). The images also display a 'halo' effect around the growing crystals which is attributed to the presence of depletion zones (i.e. solution regions which are depleted of this coloured protein). The crystals from the USML-2 mission (which were not monitored by CCD video) underwent an X-ray mosaicity analysis. The quality enhancement between Earth-grown and microgravity-grown crystals was not as marked as for the lysozyme crystals grown in dialysis geometry, nor was it consistent between the two populations, although the best crystal was microgravity grown (Snell *et al.* 1997*a*).

(b) *Lysozyme on the SpaceHab-01 mission*

WFOV CCD video monitored lysozyme crystallization in one experimental reactor onboard SpaceHab-01. Although the images do not have a particularly high sampling frequency (about 8 h between each image), and a very detailed examination has not taken place, it was notable that a bulk sedimentation of crystals occurred at the time the EURECA satellite was retrieved. The gravitational acceleration at that moment onboard the shuttle was in excess of 1300 µg, i.e. mg. X-ray mosaicity analyses of the crystals grown on this mission showed the most perfect protein lipids ever seen (minimum mosaicity observed was 0.001°). An approximate three times improvement in crystal perfection over the ground controls was seen (Snell *et al.* 1995; Helliwell *et al.* 1996) consistent also with the lysozyme grown on the IML-2 mission (described in the next section). Riès-Kautt *et al.* (1997) state that lysozyme crystals grown on the ground and in microgravity show no significant differences in rocking curve; their minimum observed mosaicity was 0.006°.

(c) *Lysozyme on the IML-2 mission*

In the IML-2 mission one crystallization experiment was monitored using NFOV optics. A series of nine images at different focal lengths were taken over a 40 min period, followed by an 8 h wait until the next series of images was taken. We were able to conduct two CCD video analyses from these data (Snell *et al.* 1997*b*). First, a crystal nucleated, probably attached to the chamber wall, within the field of view. Analysis of this crystal revealed spurts and lulls in its growth rate (figure 2). The times of the spurts in crystal growth rate directly correlated with those of astronaut exercise periods. Analysis of gravitational accelerations onboard the orbiter showed

(a)

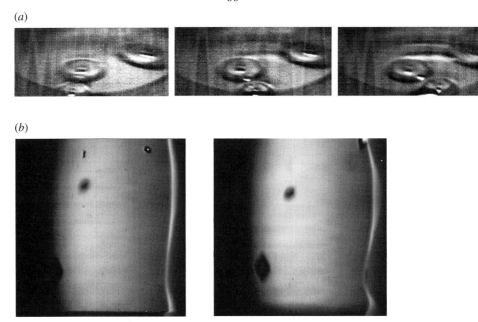

(b)

Figure 1. Examples are given from different microgravity missions in which we have been involved. (a) Cyclic movements of apocrustacyanin C_1 crystals monitored by CCD video on the IML-2 shuttle mission, using a vapour diffusion geometry in the APCF, shown here over a period of 9 min. (b) Near linear drift movements of lysozyme crystals monitored by CCD video on the IML-2 shuttle mission, using a liquid–liquid dialysis geometry onboard the APCF. Time between these two frames is 36 min.

that astronaut exercise periods, especially the use of an ergometer (a bicycle-type device), produced periods of g-jitter approaching 1000 µg. It seems that the increased gravity of these periods induced movements of fluids within the crystallization chamber, thus transporting new protein to the growing crystal faces. In our second video analysis three crystals free floating in solution were tracked over 40 min periods, between the first and last focal length images (figure 1b). The speeds of the crystal movements were seen to be of the order of $200\ \mu\mathrm{m\ h^{-1}}$, and in the same direction for all three crystals, covering a total distance of *ca.* 0.3 mm in *ca.* 40 min (i.e. the order of one crystal width). Table 1 compares such experimentally measured speeds of crystals and distances travelled. X-ray mosaicity analysis of the crystals grown on this mission showed a three times improvement in crystal perfection over the ground controls, although none of the crystals were as perfect as those from the shorter SpaceHab-01 mission, but which were also three times more perfect than their Earth-grown controls (Snell *et al.* 1995; Helliwell *et al.* 1996).

3. The EURECA uncrewed satellite and α-crustacyanin protein crystal growth

We have also analysed protein crystal growth onboard an uncrewed platform, whereby α-crustacyanin was grown on the ESA's EURECA satellite using a liquid–liquid free interface diffusion geometry. The apparatus and experiment is described elsewhere (Zagalsky *et al.* 1995; Snyder *et al.* 1991; Schmidt *et al.* 1992). We noted that for long periods of time (7 weeks) the free floating crystals nucleated and then

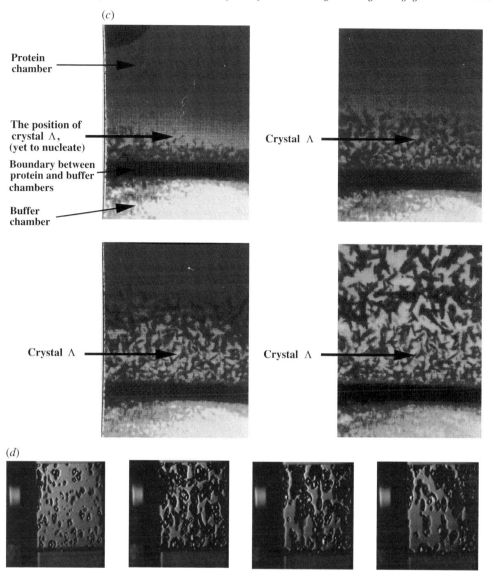

Figure 1. *Cont.* (*c*) Video images of α-crustacyanin crystal growth, in free interface liquid–liquid geometry, onboard the uncrewed satellite EURECA. The top two frames and the one at bottom left span 38 h and the one at bottom right was 26 days later. Λ indicates an easily recognized crystal that is seen to remain stationary throughout the period shown. *Stationary* crystal growth is seen over a period of 7 weeks in fact. (*d*) Near linear drift movements of lysozyme crystals monitored by CCD video on the LMS shuttle mission, using a liquid–liquid dialysis geometry onboard the APCF (see also figure 5). The crystal drift is towards the top of the reactor and from left to right. These four frames span the 15 days of the mission.

grew without moving (Boggon *et al.* 1998) (figure 1*c*), and were only finally disturbed by failures of the temperature control system. Analysis of the gravitational acceleration environment showed that a maximum of only 62.5 µg (Eilers & Stark 1993) was experienced onboard the uncrewed platform. This maximum gravitational acceleration was an order of magnitude lower than that experienced by crystal-growth

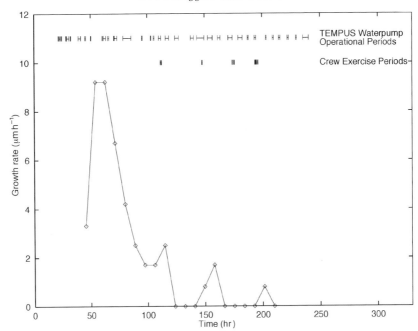

Figure 2. Spurts and lulls in growth show a correlation with astronaut exercise periods (*ca.* 5 Hz, plus harmonics, *g*-jitter) presumably inducing fluid flow and revived transport of protein to the growing protein crystal. These are possible explanations of the mosaic blocks seen in topographic images (see Chayen *et al.* (1996) and figure 7).

experiments onboard the crewed Space Shuttle described above. The spatial stability of the crystals grown onboard EURECA corresponds then to a spatial coherence length of at least as small as 1–2 CCD pixels (*ca.* 25–50 μm) (and probably better than this) for the fluid over 7 weeks. This is clearly the degree of crystallization stability, of the crystals and fluid, that can be aimed for. Indeed, the crystals grown on this mission were bigger, and had a better morphology than seen on Earth before (Zagalsky *et al.* 1995). It was most unfortunate that the experimental temperature regulation apparatus failed and temperatures as high as 40 °C were experienced by the crystals.

4. Crystal growth in gels

Crystallization in gels can suppress fluid and crystal movement following nucleation. Moreover, the rate of growth can be reduced and might then generally lead to larger, higher-quality crystals of enhanced stability. Presumably these effects could arise from the decreased mobility of the macromolecules and their flux at the crystal surface during growth, similar to microgravity.

In particular, by reducing or eliminating density-driven and Marangoni convective flow patterns, a more controlled environment may be generated around crystal-growth surfaces. The sedimentation of nucleated crystals is minimized within the gel matrix. Interestingly (figure 3) the number of crystal nucleation sites may be reduced in gels, depending on the gel material (Provost & Robert 1991). Our gel-grown crystals are grown as follows. The gel used was a physical hydrogel of agarose at a concentration 0.2%. The hen egg-white lysozyme concentration was 25 mg ml^{-1} in the buffer, which consisted of sodium acetate and acetic acid (pH 4.5); the salt

Table 1. *Experimental speeds and distances travelled of microgravity-grown protein crystals for the cases shown in figure 1*

(FID, liquid–liquid free interface diffusion; DIA, liquid–liquid dialysis; VAD, vapour diffusion (hanging drop).)

mission	protein	crystallization geometry	time period	speed $(\mu m\ s^{-1})$	total observed movement
IML-2	apocrustacyanin C_1	VAD	8 min 45 s	$2.1^\S$	1.08 mm
IML-2	lysozyme	DIA	*ca.* 40 min periods	0.05	0.3 mm
EURECA	α-crustacyanin	FID	7 weeks	*ca.* $0^\dagger$	*ca.* $0^\dagger$
LMS	lysozyme	DIA	24–$260\ h^\ddagger$	0.004^*	0.42–$3.38\ mm^\ddagger$
			ca. 2 h[a]	0.026^*	0.21 mm
			ca. 2 h[b]	0.027^*	0.17 mm
			ca. 2 h[c]	0.031^*	0.19 mm

$\S$This observation compares well with theoretical estimates (table 2).

$\dagger$The movement of crystals, for a 7 week period was less than 1–2 pixels (i.e. less than 25–50 μm); thereafter problems with the cooling elements caused crystal movements.

$^*0.004\ \mu m\ s^{-1} = 40\ \text{Å}\ s^{-1}$.

$\ddagger$For each crystal analysed as shown in figure 5.

*Instantaneous speed for all crystals over the time periods: (a), MET 3/02:35–3/04:47; (b), MET 4/04:48–4/06:31; (c), MET 6/01:18–6/03:01.

Table 2. *Theoretical estimates for lysozyme crystal speeds for vapour diffusion (Savino & Monti 1996)*

location	geometry	speed $(\mu m\ s^{-1})$
Earth	hanging droplet ('half')	5
Earth	sitting droplet ('half')	15
$0g$	'full' droplet	500

solution consisted of sodium chloride. A small amount of MPD was added to the protein solution. The glass tube used was 6 cm tall with an inner diameter of 4 mm. These crystals await detailed mosaicity, topography and resolution limit testing at the time of writing. Other gel crystal-growth experiments have been performed on acidic phospholipase A_2 in attempts to mimic the environment of microgravity (Bi 1997).

5. Recent LMS mission results

Four lysozyme crystallizations were carried out in microgravity on the LMS mission, and four identical experiments were conducted as ground controls during the period of the LMS mission. The ground control crystallizations were set up at the same time, and using the same solutions, as those of the mission. All parameters

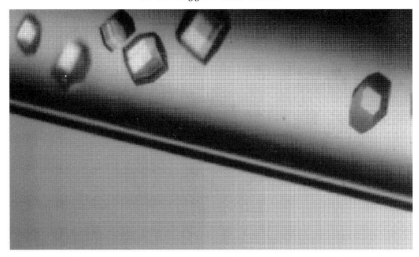

Figure 3. Gel-grown hen egg-white lysozyme crystals.

were kept as identical as possible to the microgravity case, although a temperature stability during growth of $20 \pm 0.1\,^{\circ}\mathrm{C}$ was not possible, and $\pm 1\,^{\circ}\mathrm{C}$ was used. Two crystallization conditions, one with a higher precipitant concentration, and one with a 7% lower precipitant concentration were used. Two microgravity reactors were monitored using Mach–Zehnder interferometry and CCD video, the other two microgravity reactors could not be visually monitored during the mission. All reactors produced crystals. A variety of X-ray evaluations were conducted on these crystals.

(a) CCD video

On the LMS mission two reactors were monitored using CCD video, they were numbered reactors 7 and 9 in the APCF. For reactor 7, WFOV images of the whole protein chamber were taken at one focal length position every 2 h, beginning 15 h after activation of the APCF. In reactor 9, the first image was taken 90 h after APCF activation. CCD video analysis of these two reactors complemented Mach–Zehnder interferometry analysis (§ 5 *b*).

The monitoring of reactor 7 showed that nucleation events occurred depending on the spatial position within the reactor, with the first crystals visible 25 h 37 min after APCF activation close to the dialysis membrane (figure 4). We noted significant crystal 'drifting' movements (figure 5) throughout the mission, whereby nucleation events occurred throughout the chamber but were very quickly followed by a global sedimentation drift towards the top of the reactor. The only crystals that remained in their nucleation positions were those that grew on the chamber walls. By following 20 different crystals for the periods that they could be accurately tracked sudden global movements could also be seen, in addition to the steady drift, at the 76, 102 and 145 h points, as well as other times. The source of the sudden disturbance at these times into the LMS mission is being investigated. The total drift distance, for all 20 crystals analysed, over the periods they were analysed, ranged between 0.42 and 3.38 mm, with an average speed of *ca.* $0.004\,\mu\mathrm{m\,s^{-1}}$ ($40\,\text{Å s}^{-1}$).

(b) Mach–Zehnder interferometry

On the LMS mission, the two reactors described above were also monitored using Mach–Zehnder interferometry. Reactor 7 contained a slightly higher precipitant con-

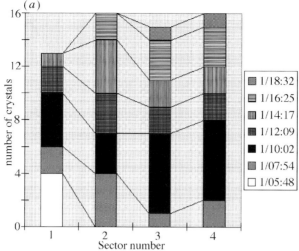

Figure 4. (*a*) The dependence of spatial position of nucleation of crystals. The different shades of the bars correspond to the number of crystals that have nucleated by that time (see the key in days/hours: minutes) within each of the four sectors.

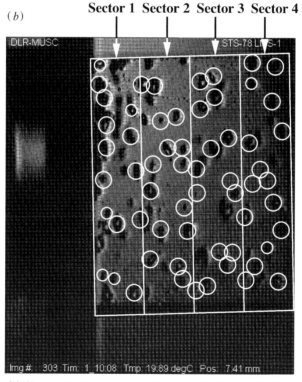

Figure 4. *Cont.* (*b*) Shows the four sectors referred to in (*a*) within the reactor with the crystals monitored circled.

centration and so it was monitored at a slightly faster rate to compensate for the expected increased speed of equilibration of precipitant. Due to time constraints on digitally recording the images (as with the CCD video, 4 min per whole image), and

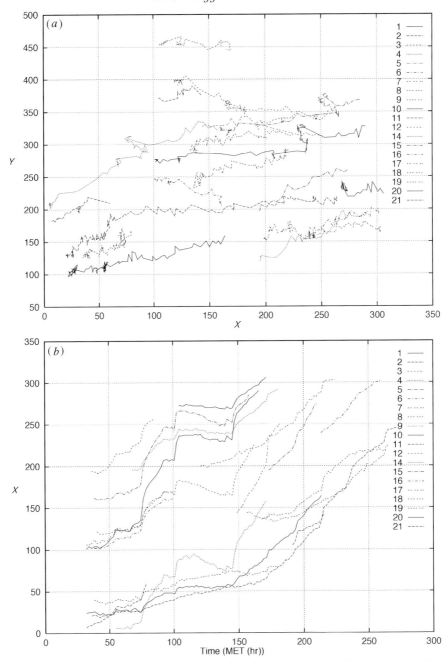

Figure 5. Crystal movement throughout the LMS mission in reactor 7. Three graphs illustrating the movements of 20 crystals are shown. Graph (*a*) shows X versus Y, (*b*) X versus time. The units are pixels and hours (mission elapsed time, MET). The positions of crystals in figure 5*a* can be overlaid onto figure 1*d*.

the need to accurately follow the progression and number of fringes, whole interferograms could not be collected for the first 15 h after APCF activation. Instead two windows were defined (w1 and w2 in figure 6*a*), which monitored the local average

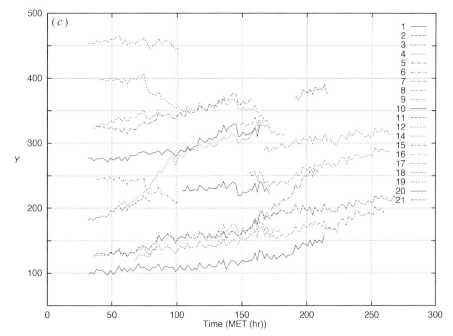

Figure 5. *Cont.* (*c*) *Y* versus time.

grey value at 4 or 6 min intervals for reactors 7 and 9, respectively. After 15 h, whole reactor interferograms were collected at 11 min intervals for both reactors.

Unfortunately the laser was unsteady, and mode switching occurred throughout the mission. This was problematic, especially in the first 15 h, when average grey values were the only method of analysis. Consequently, some fringes were probably missed in that period. However, a more complete analysis was possible with the whole interferograms collected after the 15 h point, although these also have been affected by the problematic laser (figure 6).

There was, however, enough data to yield some very interesting results. First, the direction of fringe movement reversed at about the 8 h point in both reactors, and then again at the point that crystals became visible in the CCD video monitoring, i.e. between 25 and 31 h (depending on the position within the protein chamber) (figure 6*e*). Compared to the solitary reversal in fringe movement direction present in the ground control experiments (Snell *et al.* 1996), this is a surprising result which suggests a different pre- and post-nucleation process occurs in microgravity compared to the Earth-grown case. This must be looked at further.

Finally, the whole chamber interferograms demonstrated the stability of fluid within microgravity crystallization by showing an effect that is rarely seen using Mach–Zehnder interferometry on Earth, namely depletion zones are clearly visible around growing crystals (figure 6*b, d*).

(*c*) *Crystal perfection assessment using X-ray reciprocal space mapping and topographic techniques*

Synchrotron X-ray analysis was carried out at the NSLS using beamline X26C on the LMS crystals. Protein crystal mosaic spread measurements by use of rocking curves is an indicator of the internal physical perfection present (Helliwell 1988). The addition of an analyser crystal between the sample and detector enables recip-

(a)										(b)

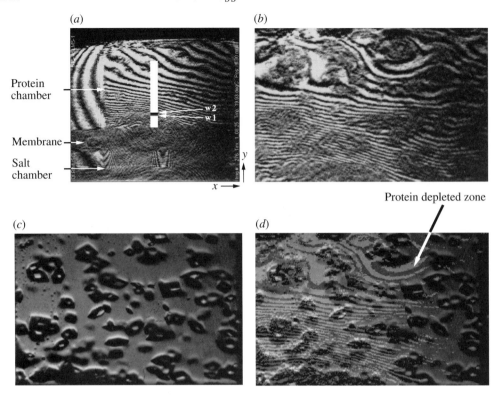

Protein chamber

Membrane

Salt chamber

Protein depleted zone

(c)										(d)

Figure 6. Diffusion of precipitating agent into the protein chamber of a crystallization reactor and subsequent 'removal' of protein into nucleating and then growing crystals can be monitored via Mach–Zehnder interferometry. (a) An example of a whole interferogram with interference fringes clearly visible. The two grey value measuring windows are marked as w1 and w2, and the white strip indicates the area analysed in (e). (b) A fringe image, later into the crystallization. (c) A CCD video image taken at the same time as the fringe image (b). (d) Overlay of (b) and (c) showing a protein depletion zone (marked).

rocal space mapping of ω, the sample axis, and ω', the analyser axis (Snell 1998). Reciprocal space maps along one axis provide a measure of pure mosaicity effects (volume and orientation), and along the other axis strain effects. Combination of the use of this technique with X-ray topography (Stojanoff *et al.* 1996, 1997), can produce a finely detailed picture of a single reflection, and an in-depth knowledge of the internal order of the crystal. Two examples are shown here (figure 7).

(d) The X-ray data collection from the LMS and ground control lysozyme and apocrustacyanin C_1 crystals, under identical laboratory X-ray source conditions

Complete X-ray single crystal diffraction data sets were measured on one LMS lysozyme and one Earth control crystal sample. Details of the data processing are given in table 3. Table 4 is a similar analysis for apocrustacyanin C_1, but where a large difference in crystal volume is the predominant effect. The procedures followed for data collection were essentially identical for the microgravity and Earth lysozyme cases. The data quality from the microgravity and Earth control lysozyme crystals is essentially identical using coarse angle rotation ranges. For the LMS microgravity-grown lysozyme crystal the data collection protocol was as follows: 60° of 0.75°

(*e*)

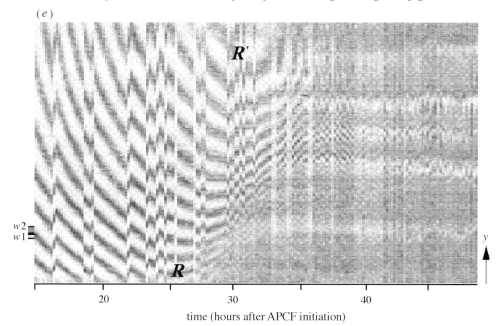

Figure 6. *Cont.* (*e*) Analysis of the movement of fringes within the white strip shown in figure 6*a*. The average grey value of pixels in *x* within the white strip has been obtained and shown here as a single pixel. The height (*y*) is unchanged for both sets of pixels.

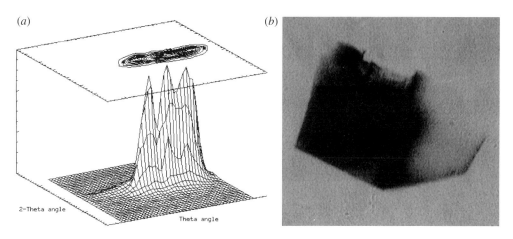

Figure 7. Protein crystal quality assessment via (*a*) reciprocal space mapping and (*b*) topographic images for lysozyme crystals grown aboard the LMS shuttle mission as examples.

oscillations at a distance of 250 mm using 30 min exposures, followed by 60° of 0.50° oscillations at a distance of 300 mm using 20 min exposures. For the Earth-grown crystal the data collection protocol was identical, and then in addition 60° of 1.5° oscillations at a distance of 450 mm using 30 min exposures were recorded. The LMS crystal was $0.50 \times 0.50 \times 0.35$ mm^3 (0.0875 mm^3) and the Earth control crystal was $0.70 \times 0.50 \times 0.45$ mm^3 (0.1575 mm^3).

Resolution limits are dependent on the data collection source and measuring conditions. In earlier work on lysozyme, use of the Swiss–Norwegian beamline at the

Table 3. *X-ray data sets collected on LMS and LMS ground control lysozyme crystals on the Manchester R-AXIS IIc IP area detector MoKα ($\lambda = 0.71$ Å wavelength)*

(Overall I/Sigma I for LMS is 15.7, for LMS ground control is 16.5. Crystal volume for LMS is 0.0875 mm^3, and for LMS ground control is 0.1575 mm^3.)

	LMS		LMS ground control	
resolution	R_{merge}	completeness %	R_{merge}	completeness %
99.00–4.19	0.054	95.1	0.056	98.8
4.19–3.33	0.064	98.8	0.064	99.2
3.33–2.91	0.075	99.2	0.070	97.1
2.91–2.64	0.085	99.6	0.080	97.5
2.64–2.45	0.093	99.9	0.090	96.6
2.45–2.31	0.093	99.6	0.090	97.7
2.31–2.19	0.099	99.3	0.097	96.9
2.19–2.10	0.105	98.1	0.105	96.1
2.10–2.02	0.120	94.3	0.116	94.9
2.02–1.95	0.135	88.8	0.139	92.1
1.95–1.89	0.165	77.3	0.163	88.7
1.89–1.83	0.186	66.5	0.187	81.4
1.83–1.78	0.239	62.1	0.249	70.9
1.78–1.74	0.271	54.5	0.255	64.3
1.74–1.70	0.393	45.1	0.344	56.1
overall	0.072	85.5	0.069	88.8

ESRF and an extra fine ω step diffractometer ($\delta\omega \approx 10^{-4}$ deg) yielded reflections at 1.2 Å for microgravity lysozyme but not the Earth-grown crystals (see figures 3 and 4 of Snell *et al.* (1995)). This can be contrasted with the resolution limit here for laboratory data sets of LMS lysozyme crystals of *ca.* 1.7 Å. The full exploitation of the crystal perfection available is very difficult if step widths of 10^{-4} degrees are required. Nevertheless, new area detectors like the pixel detector could exploit such quality (Chayen *et al.* 1996) when used in conjunction with X-ray undulator sources, which have extremely fine collimation in horizontal *and* vertical directions. (For a discussion of detector and synchrotron radiation sources for macromolecular crystallography see Helliwell (1992).) It is the geometric quality and internal perfection as well as the sizes of protein crystals, which multiplied together (i.e. mosaicity × sample size) form the parameter (sample acceptance) that is comparable directly with synchrotron machine emittance. Earlier work comparing mosaicity of microgravity and Earth-grown crystals of lysozyme and apocrustacyanin C_1 can be found in Helliwell *et al.* (1996) and Snell *et al.* (1995, 1997a). Examples of topographs of Earth- and microgravity-grown lysozyme can be found in Stojanoff *et al.* (1996) and Chayen *et al.* (1996).

6. Conclusions

Microgravity protein crystal growth can enable the crystallization process to proceed in a relatively undisturbed state, with diffusion dominating rather than fluid

Table 4. *Resolution limits of apocrustacyanin C_1 crystals grown onboard the LMS missions in a vapour diffusion reactor and Earth-grown controls determined from $1°$ oscillation images using ESRF BL19.*

resolution	LMS		LMS ground control	
	R_{merge}	completeness %	R_{merge}	completeness %
25.00–4.92	0.048	86.5	0.045	97.2
4.92–3.91	0.048	91.0	0.047	99.0
3.91–3.42	0.050	92.6	0.070	99.5
3.42–3.11	0.051	94.2	0.089	99.8
3.11–2.88	0.058	94.7	0.130	99.9
2.88–2.71	0.064	95.0	0.172	99.7
2.71–2.58	0.070	95.3	0.203	99.8
2.58–2.47	0.074	96.6	0.230	99.6
2.47–2.37	0.075	96.1	0.239	99.9
2.37–2.29	0.084	95.2	0.266	99.5
2.29–2.22	0.084	97.9	0.265	99.7
2.22–2.15	0.085	95.7	0.275	99.1
2.15–2.10	0.086	98.1	0.317	98.6
2.10–2.05	0.094	95.3	0.358	98.9
2.05–2.00	0.104	98.2	0.403	98.2
Overall	0.056	94.7	0.110	99.2

ESRF BL19 is a bending magnet source beamline. Operating parameters during data collection were $\lambda = 0.7513$ Å; CCD detector (image intensifier type), exposure times 30 s, rotation angles $1°$, $\delta\lambda/\lambda \sim 10^{-4}$. Overall I/Sigma I for LMS is 30.4, and for ground control is 9.5. LMS crystal volume is 0.0432 mm^3, ground control crystal volume is 0.0034 mm^3.

Data collection of an Earth-grown crystal on ESRF BL4 gave diffraction to 1.3 Å. The crystal's dimensions were $0.2 \times 1.25 \times 0.5$ mm^3 (0.125 mm^3). ESRF BL4 is fed by an X-ray undulator source (horizontal and vertical divergences of the beam *ca.* $0.005°$, i.e. 0.1 mrad). Detector used Mar IP, exposure times 30 s, rotation angles $1°$, $\lambda = 0.94$ Å, $\delta\lambda/\lambda \sim 10^{-4}$.

The positive influence of a stronger beam, and a bigger crystal volume is illustrated by the effects on the resolution limits of these crystals, i.e. in the main table 4, as well as the BL4 test.

flow. The closest to this ideal situation has come from the use of a linear liquid–liquid free interface diffusion crystallization geometry and on an uncrewed platform in microgravity (EURECA). Use of vapour diffusion hanging drops is prone to crystal movements which have the form of Marangoni stimulated fluid flows. It would be a pity if hanging drops could not be used in microgravity, due to Marangoni fluid flow, since it is a method rather economical of protein. Fortunately, there are other micromethods that might be tried in microgravity, which avoid the liquid–vapour interface, and thereby Marangoni convection. In addition, *g*-jitter events (especially at low frequencies) appear to stimulate crystal movements and also fluid flows. The overall drift speed of lysozyme crystals on the LMS shuttle mission of 40 Å s^{-1} presumably is tolerable but is of interest in modelling the biophysical chemistry of protein crystal growth. However, the cases of sudden jumps on LMS of, for example, *ca.* 0.2 mm over 2 h periods (i.e. 0.03 µm s^{-1} or 300 Å s^{-1}), due presumably to *g*-jitter disturbances, is a more serious concern for realizing the best quality protein crystals. Nevertheless, improvements in crystal perfection have been observed on a number of microgravity shuttle missions and which are coming close to the theoretical limits of the rocking curve and spatial coherence of a protein crystal. The mimicking of microgravity, and thereby the matching on Earth of microgravity benchmark levels of quality is a key target; perhaps gel-based growth will provide these matching benchmarks. The full exploitation of such quality in X-ray crystallography is a challenge, and an opportunity, for computing and 'diffractometer' area-detector hardware. Such studies also require improved sample freezing techniques; the latter are essential to control X-ray damage to, but lead to less geometrical perfection of, the sample. A possible additional physical process for improved 'Bragg' diffraction data for microgravity-grown crystals is that disorder diffuse scattering (DDS) may be reduced. This has been seen in chemical crystallography for microgravity- versus Earth-grown crystals (Ahari *et al.* 1997), and may apply to protein crystals thus explaining why only a fraction of proteins and protein crystal types appear to benefit from microgravity (i.e. DDS is not commonly prevalent). Overall, improved crystal order via better crystal growth, and the understanding of the fluid physics which underpins it, will stimulate new developments in X-ray data collection and thereby data quality. Moreover, protein structure projects where only microcrystals can be grown, or perhaps projects with no crystals at all, might also benefit from these new developments in fluid physics and 'materials processing'.

Dr J. Stapelmann, R. Bosch, W. Fritzsch and W. Scheller of Dornier GmbH are thanked for their continuing help and support with this work and F. Bottcher for technical assistance. Dr H. U. Walter, Dr K. Fuhrmann, Dr H. Martinides and Dr O. Minster at ESA are acknowledged for their support and the opportunity to perform these experiments. The NSLS is acknowledged for providing beam time for the analysis of crystal perfection. The NSLS is funded by the US Department of Energy under contract number DE-AC02-76CH00016. Likewise the ESRF is thanked for provision of synchrotron radiation on BL19 and the undulator line BL4. The Wellcome Trust funded the Manchester R-AXIS to whom J.R.H. is grateful. J.D. was supported by the Royal Society for a one-year QE fellowship in Manchester on leave from Beijing. T.J.B. thanks the University of Manchester and the Samuel Hall fund for studentship support. The SRS is thanked for provision of synchrotron radiation on station 9.5 for ultra-long distance Laue mosaicity evaluation, as well as LURE and at the ESRF (Swiss–Norwegian beamline) for rocking curve measurement runs, results from which are described in the abstract.

References

Ahari, H., Bedard, R. L., Bowes, C. L., Coombs, N., Dag, Ö., Jiang, T., Ozin, G. A., Petrov, S., Sokolov, I., Verma, A., Vovk, G. & Young, D. 1997 Effect of microgravity on the crystallization of a self-assembling layered material. *Nature* **388**, 857–860.

Bi, R.-C. 1997 Protein crystal growth in space. In *Space science in China* (ed. W.-R. Hu), pp. 397–413. London: Gordon and Breach.

Boggon, T. J., Chayen, N. E., Zagalsky, P. F., Snell, E. H. & Helliwell, J. R. 1998 Stationary crystal growth of α-crustacyanin in microgravity onboard the unmanned EURECA carrier, monitored by video observation. *Acta Crystallogr.* D. (Submitted.)

Bosch, R., Lautenschlager, P., Potthast, L. & Stapelmann, J. 1992 Experimental equipment for protein crystallization in microgravity facilities. *J. Cryst. Growth* **122**, 310–316.

Chayen, N. E., Boggon, T. J., Cassetta, A., Deacon, A., Gleichmann, T., Habash, J., Harrop, S. J., Helliwell, J. R., Nieh, Y. P., Peterson, M. R., Raftery, J., Snell, E. H., Hädener, A., Niemann, A. C., Siddons, D. P., Stojanoff, V., Thompson, A. W., Ursby, T. & Wulff, M. 1996 Trends and challenges in experimental macromolecular crystallography. *Q. Rev. Biophys.* **29**, 227–278.

Chayen, N. E., Snell, E. H., Helliwell, J. R. & Zagalsky, P. F. 1997 CCD video observation of microgravity: apocrustacyanin C_1. *J. Cryst. Growth* **171**, 219–225.

Eilers, D. & Stark, H. R. 1993 EURECA microgravity environment: preliminary flight data. *NASA Conf. Proc.* **3272**, 869–891.

Helliwell, J. R. 1988 Protein crystal perfection and the nature of radiation damage. *J. Cryst. Growth*, **90**, 259–272.

Helliwell, J. R. 1992 *Macromolecular crystallography with synchrotron radiation.* Cambridge University Press.

Helliwell, J. R., Snell, E. H. & Weisgerber, S. 1996 An investigation of the perfection of lysozyme protein crystals grown in microgravity and on Earth. In *Proc. 1995 Berlin Microgravity Conf.* (ed. L. Ratke, H. U. Walter & B. Feuerbacher), pp. 155–170. Berlin: Springer.

Provost, K., & Robert, M. C. 1991 Application of gel growth to hanging drop technique. *J. Cryst. Growth* **110**, 258–264.

Riès-Kautt, M., Broutin, I., Ducruix, A., Shepard, W., Kahn, R., Chayen, N.E., Blow, D., Paal, K., Littke, W., Lorber, B., Théobald-Dietrich, A. & Giegé, R. 1997 Crystallogenesis studies in microgravity with the advanced protein crystallization facility on SpaceHab-01. *J. Cryst. Growth* **181**, 79–96.

Savino, R. & Monti, R. 1996 Buoyancy and surface-tension-driven convection in hanging-drop protein crystallizer. *J. Cryst. Growth* **165**, 308–318.

Schmidt, H. P., Koerver W. & Pätz, B. 1992 Practical aspects of crystal growth experiments on board EURECA-1. *J. Cryst. Growth* **122**, 317–322.

Snell, E. H. 1998 Quality evaluation of macromolecular crystals using X-ray mosaicity measurements. *Proc. Montreal Spacebound 97 Meeting, Canadian Space Agency.* (In the press.)

Snell, E. H., Weisgerber, S., Helliwell, J. R., Weckert, E., Hölzer, K. & Schroer, K. 1995 Improvements in lysozyme protein crystal perfection through microgravity growth. *Acta Crystallogr.* D **51**, 1099–1102.

Snell, E. H., Helliwell, J. R., Boggon, T. J., Lautenschlager, P. & Potthast, L. 1996 Lysozyme crystal growth kinetics monitored using a Mach–Zehnder interferometer. *Acta Crystallogr.* D **52**, 529–533.

Snell, E. H., Cassetta, A., Helliwell, J. R., Boggon, T. J., Chayen, N. E., Weckert, E., Hölzer, K., Schroer, K., Gordon, E. J. & Zagalsky, P. F. 1997*a* Partial improvement of crystal quality for microgravity grown apocrustacyanin C_1. *Acta Crystallogr.* D **53**, 231–239.

Snell, E. H., Boggon, T. J., Helliwell, J. R., Moskowitz M. E. & Nadarajah, A. 1997*b* CCD video observation of microgravity crystallization of lysozyme and correlation with accelerometer data. *Acta Crystallogr.* D **53**, 747–755.

Snyder, R. S., Fuhrmann, K. & Walter, H. U. 1991 Protein crystallization facilities for microgravity experiments. *J. Cryst. Growth* **110**, 333–338.

Stojanoff, V., Siddons, D. P., Snell, E. H. & Helliwell, J. R. 1996 X-ray topography: an old technique with a new application. *Synchrotron Radiation News* **9**, 25–26.

Stojanoff, V., Siddons, D. P., Monaco, L. A., Vekilov, P. G. & Rosenberger, F. 1997 X-ray topography of tetragonal lysozyme grown by the temperature controlled technique. *Acta. Crystallogr.* D **53**, 588–595.

Zagalsky, P. F., Wright, C. E. & Parsons, M. 1995 Crystallisation of α-crustacyanin, the lobster carapace astaxanthin-protein: results from EURECA. *Adv. Space Res.* **16**, 91–94.

Author Index

Subject Index